I0796579

THE ADVENTURES OF A NARRATIVE GARDENER

THE ADVENTURES OF A NARRATIVE GARDENER

Creating a Landscape of Memory

Ronald Lee Fleming

The Townscape Institute
In association with D Giles Limited

This book is dedicated to my parents and my great-grandfather Edmund Wilson Reynolds, as stewards who gave me the confidence to find my own voice and the means to achieve—and to my children, Severine, Siena, and Reynolds, whose eyes keep the world fresh and warm and whose spirits sustain their father. They are the ultimate anchors of place.

First published in 2020 by GILES
An imprint of D Giles Limited
66 High Street,
Lewes, BN7 1XG, UK
gilesltd.com

ISBN: 978-1-911282-74-7

For the Townscape Institute:
Project Director: Ronald Lee Fleming
Production Manager: Matthew Esposito

For D Giles Limited:
Copy-edited and proof-read by Jodi Simpson
Designed by Caroline and Roger Hillier,
The Old Chapel Graphic Design
Produced by GILES, an imprint of D Giles Limited
Printed and bound in Slovenia

Front cover: View from the Oriental Vale to the Samuel McIntire-inspired rotunda, based on his design for the Howard Street Church in Salem, Massachusetts
Back cover: Chinese Chippendale bridge, Bellevue House
Frontispiece: View east to the rotunda from centerpoint of the hornbeam allée

Contents

Acknowledgments 6
Foreword 14

Chapter One
NAVIGATING THE COSMOS OF FAMILY
Piecing together the landmarks that define it and the patterns that connect it
25

Chapter Two
THE PLACE OF MEMORY
Resonating in the soul
63

Chapter Three
THE PLACE OF ANIMATION
Expressing the spirit
85

Chapter Four
MEANING AND MEMORY COME TOGETHER IN THE ACADEMICAL VILLAGE
101

Chapter Five
THE PLACE OF CRAFT AND THE ROLE OF THE ARTIST
119

Chapter Six
THE PLACE OF HUMOR
139

Artists and artisans 146
Appendices 151
Historic gardens 154
Selected bibliography 157
Index 163
Photography credits 168

Acknowledgments

Because this book tells a story about family and place, it necessarily chronicles the nourishment of a vision, the focus of which is Bellevue House and its garden, in Newport, Rhode Island. This vision was empowered by the growth of imagination. At the age of four, in a musty beach-house basement, I found the neatly catalogued postcards that my grandparents had left behind from their round-the-world voyage on the P&O Line in 1922 and explored the artifacts of that journey, from amulets of Jordan River water to Philippine knives to bits of marble from the Parthenon. Probably I was sensitized by these early influences. Maybe it was environmental determinism, but perhaps there was a rough-hewn sensibility, inchoate but embedded within, that shaped this eventual narrative gardener.

I grew from an inquisitive child to an inquiring adult, viewing old graveyards, visiting ghost towns, collecting rocks, stamps, and weapons, and later collecting stories. I have been privileged to hear many stories over my lifetime, like the one heard in a plantation graveyard in Orange, Virginia: the beloved maiden aunts of my hostess's grandmother wouldn't leave the kitchen to attend the parlor wedding of their niece who was marrying a Yankee lawyer she had met while visiting New York.[1] I saw how the bitterness of the past expressed itself in the stifled lives

Fig. 1 Exterior view of the Berkeley Villa/Martha Codman Karolik House, now Bellevue House, Newport, Rhode Island, c. 1910–38

1

of these old maids whose potential lovers were killed in our most costly war. I caught the sense of lost time, of a deeper culture where a sense of a shared past has disappeared in the face of suburban isolation, and where the old-fashioned idea of personal honor has been largely eroded.

Thus, I pause now to pay tribute to those who have shaped my imagination. This book is probably the creative culmination of my career. It has been about twelve years since my last publication, *The Art of Placemaking.* The present book aims to bring that art into the garden, an often ephemeral and fragile space. This book has been more time-consuming than my previous volumes, which featured projects that I or others have completed. As in my garden, the work seemed never-ending, as the eye measured what still needed to be done, and done better.

I begin my acknowledgments by commemorating a fifteen-year collaboration with J. P. Couture, who is both architect and preservation planner, just as I have been both placemaker and artistic visionary. Somehow, we have prevailed in our projects without ever throwing an eraser at one another; usually we shared photographs of the buildings and gardens we liked from around the world. This collaboration has enriched our lives, even as it has depleted my pocketbook. I am grateful that the garden spaces have been populated with follies and memories that we can now celebrate in this book.

Glenn Parker, our general contractor, supervised most construction projects. As a former doctoral candidate sensitive to design quality, he enabled us to use two photographs from the documentation he commissioned from Warren Jaeger. He personally intervened to secure

Fig. 2 Bellevue House, c. 2018. Minimal changes have been made to the integrity of architecture and design. New yews have been planted in jardinières and flowers in urns as a welcoming gesture to guests.

2

3

delivery of the lotus carving made by English stone-carver Nicholas Fairplay, to evoke the idea of fertility rising out of the muck. Edwin Lutyens designed these lotus flowers in massive brownstone for the Mughal Gardens behind the Vice Regal Lodge in New Delhi.

The next to be honored is the late Lester Glenn "Ruff" Fant III, who had a curable form of cancer but died suddenly and unexpectedly from an adverse reaction to a cancer drug in May 2019. He was one of my dearest friends and supported my vision with a most extraordinary selfless dedication. I met him at a Harvard dining club in 1964, and I am forever grateful for his advice, encouragement, and support. He was a comrade soul who will remain in my memory forever. My grief in his passing is surpassed only by my admiration of his generosity of spirit, as he shared friends, ideas, and insights.

Twelve years is a long period since the last book, and I am grateful for the patience of friends and colleagues. Of course, we must take complete responsibility for our own assessments, but there is much I would not have known without the candor of so many sophisticated observers of the

Fig. 3 Aerial view of Bellevue House and garden, 2010. This photo shows how the gardens have been transformed slowly over time. By this point, we had constructed the new small teahouse (lower left), the cupola, and the Oriental Vale garden (right).

built environment. Professional colleagues are acknowledged in the endnotes, and photographers are credited in the in the credits list at the back of the book. I also want to offer sincere thanks to those who took the time to call back to answer questions, or who had the patience to review drafts.

I am grateful to the board of the Townscape Institute, which, guided by lawyer William S. Strong, has encouraged this project over two decades. As the garden and library evolved, the late Dr. John Constable always took a renaissance view with an analytical frame of mind that has been an inspiration. He was an astonishingly loyal friend through thick and thin.

Over the years, there have been advisers and friends. Professor Nathan Glazer encouraged me to publish in the very last issue of *The Public Interest*. The resultant article, co-

4

written with Townscape staffer Melissa Tapper Goldman, helped us to press our argument about federal policy and public art.[2] Professor Glazer was a true public intellectual who didn't hoard his talent in academia. He spoke out publicly in the national press, where he could cajole, chide, and critique, even if he was sometimes "politically incorrect" in his judgments.

The present book represents a culmination of efforts over many years, and I wish to thank those individuals and families who have nourished my body and soul during the past decades, and those who were initial mentors and muses in my youth. Langley Keyes of Cambridge, Massachusetts—whose mother-in-law took me to that strawberry supper on Craftsbury Common that so influenced my vision of America (see p. 95)—encouraged me to change from the law to urban planning, which gave me an entirely different career. For that first grounding of support, I would like to thank the late Sir George and Lady Trevelyan, whom I met at Attingham Park in Shropshire, where I went to take a course on the English country house as an antidote to my experiences as an intelligence officer serving with the Fifth Special Forces Group in Nha Trang and later attached to the embassy in Saigon (see chapter 4); through a cousin of theirs, I met the late Earl Peregrine St. Germans, who enriched my vision and became a lifelong friend. The Trevelyans' daughter Catriona Trevelyan Tomalin Tyson became a good friend and, with her second husband, Richard Tyson, further opened my eyes to selected bits of the English countryside; Catriona had been the head girl at a school in Shropshire and introduced me to the successive head girl, Fiona Peel. Indeed, my Wanderjahr, which started at Attingham, included the inspection of many places, from Venice to Vienna to southern Germany and back to

Fig. 4 Aerial view of Bellevue House and garden, 2018. The landscape reveals the green house, the stable block (housing conference and guest rooms), the library/nymphaeum, and the new cabanas for visiting fellows. We had composed a new Arts and Crafts garden, the moss garden under the large canopy, new pools, and alleés. The pattern language has been enhanced with exedras, gazeboes, green garden rooms, and many volutes.

Paris, where I edited the only comprehensive collection of Czech humor from the Dubček period. This was part of a continued peregrination around Europe over many years, which has gradually increased my perception. Perhaps I will be remembered by my children for giving them each eighty days in Venice before they were twenty!

Thanks also to Frank Keefe, Marion, Massachusetts; Grant R. Jones, FASLA, a Harvard University Graduate School of Design classmate, and his wife Chong H. Jones, now of Oroville, Washington; the late Edmond H. Kellogg, Pomfret, Vermont; the late James Lawrence, FAIA, Brookline, Massachusetts; Robert and Penny McNulty of Partners for Livable Communities in Washington, D.C.; the late Professor Charles G. K. Warner, Lincoln, Massachusetts; the late Roger S. Webb, Cambridge, Massachusetts; the late William H. Whyte Jr., who gave me my first professional audience in 1971; the late Grady Clay, Louisville, Kentucky, former publisher of *Landscape Architecture Magazine*, who first published a piece I wrote about "lovable objects," which turned into a larger placemaking theory; and the late Rufus and Leslie Stillman of Litchfield, Connecticut, collectors and patrons who sheltered me and stimulated my interest in modern and contemporary art. I enjoyed staying in the Stillmans' Marcel Breuer-designed houses after my fellowship in decorative arts and history at Historic Deerfield, Massachusetts, in 1963, where I spent a blissful summer after college; indeed, reacting against modernism was part of that essential experience.

Katherine Stillman, daughter of Rufus and Leslie, and Peter Epstein, a fellow GSD classmate, actually put together the manuscript of my master's thesis on cultural planning in a new town. I had given it to a young lieutenant as I got on the plane to Vietnam after leaving Fort Bragg. To their everlasting credit and my gratitude, Peter Epstein and Katherine Stillman completed my thesis before Peter finished his own. Nore V. Winter, former director of Townscape services, now of Boulder, Colorado, and Brent C. Brolin of New York, architect and author of *Architecture in Context*, provided professional support and inspiration during the early decades of main street projects and I continue to talk with them; James Greene, architect and planner, Columbus, Ohio, continued that engagement in recent decades when I conceived placemaking projects for his firm in Centerville and Oxford, Ohio.

For their support during the past two decades, a time of stress as I sought to manage three children on my own, I should like to thank the late Lady Fiona Baker of Cambridge, England; Minette Boesel and her parents, the late William K. and Minette Bickel of Pittsburgh; the late William P. Carey, New York; Ljiljana Cook, now of New Paltz, New York; Nicholas Wellington Danforth, Weston, Massachusetts; my sister, Susan Faye Fleming, Salt Lake City; the late Ainslie Gardner, her son, Stewart, Santa Fe, and her family, Newport, Rhode Island; Rosemary and Torrance Harder, West Palm Beach, Florida; Dr. Michael Johnson, formerly of Cambridge, Massachusetts; Joanne Lawson-Derby, Washington, D.C., and Houston, Texas; and Gregg LeFevre, New York.

Fiona Peel of Cardiff, Wales, documented in a series of photobooks the progress at Bellevue House and nourished my vision. Her last little monograph, *Stately Progress at Bellevue*, was essential in recording what we have accomplished in recent years with the stable block and library/nymphaeum. Her staunch loyalty and dedication is greatly appreciated, as is that of Sandra Ourusoff, Boston, Middletown, Rhode Island, and New York; William Page Reimann, Cambridge, Massachusetts; Jill Spalding, New York; Elisabeth Vines, FRAIA, Adelaide, Australia; Rob Walker, Newport, Rhode Island; Raynor and Ranne Warner, now of Stockbridge, Massachusetts; and Kimball Wheller, Guadalajara, Mexico. For her early editing support and definition of Federalist architecture, I thank Edith Brewster, Washington, D.C., and for her painting critique, Caterine Milinaire Cushing, Newport, Rhode Island. Most recently, I received good advice and bibliographical support from Dr. Duncan Berry of West Harwich, Massachusetts.

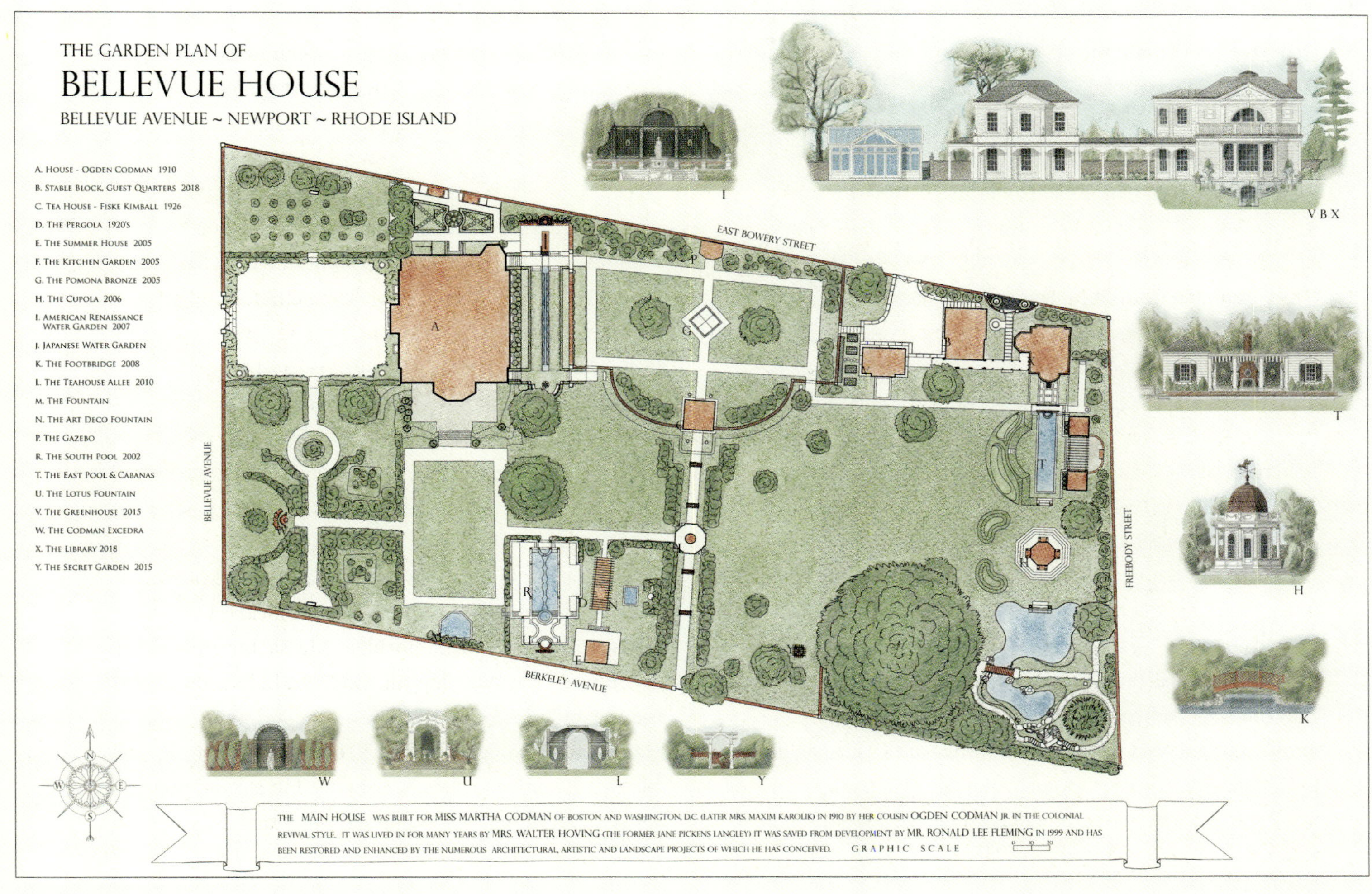

5

Fig. 5 Bellevue House garden plan, Couture Design Associates, 2019

For their assistance on the production of this book I would like to thank Matthew Esposito, Mac Griswold, the aforementioned Edith "Bobbie" Brewster, Holly Alderman and Bethy Cardoso.

Collectively, I would like to acknowledge the Diodati Dinner Club, some of whose members are listed above. I have had the pleasure of "convening" the club, modeled in part on the Society of Dilettanti in London, over three and a half decades. Its members practice the art of a rolling conversation based on the modest conceit that at least two of them can address most aspects of Western culture. Some of these comrades and friends are more thoroughly documented in chapter 4.

Finally, I would like to thank my parents, the late Mr. and Mrs. Ree Overton Fleming, formerly of Los Angeles County and then of Laguna Hills, California. To them my sister and I owe much gratitude for the happy and loving environment they provided for their children. They were the ultimate placemakers, as they secured for us a place of confidence where we could define our own challenges and later pursue the civic good on our own terms.

Taking the long view

As I was finishing these acknowledgments, John Davidson Dennis, my minister for over three score years, suddenly died in Portland, Oregon. As the longtime leader of the First Presbyterian Church of Corvallis, Oregon, he had originally brought me out to support connections between a large engineering firm moving downtown to the Willamette River edge, the university nearby, and the Madison Avenue Task Force, which he chaired at the time. We became friends because we both believed in the importance of multidisciplinary conversations between centers of power. I recognize now that this was a conviction that many of my fallen friends, memorialized on these pages, understood and shared.

Such conversations between different centers of power is one definition of a "working culture."[3] It has, after all, been a major concern. It should be the outcome of a liberal arts education. The issue is often how to build a conversation before grants or contracts are available to craft these coalitions supporting civic investment. There is evidence from late-nineteenth-century America of this engagement with particular places where local pride has inspired disproportionately large acts of community investment: for example, the railroad station in Waterbury, Connecticut, with all the grandeur of the city hall in Siena, Italy, and the courthouses proudly standing in Texas squares in communities of only modest dimension.

Today we live in an increasingly corporate state, with upwardly mobile executives focused on the bottom line. They usually are not anchored to the pride of a particular place. Jamie Gamble, a now-retired attorney, formerly of the firm Simpson, Thacher & Bartlett, who served such interests for a lifetime, recently concluded that we had turned our country over to corporate psychopaths.[4] This corporate state includes many not-for-profit organizations who also find it hard to protect a larger public interest. Often, they stay in their own lane, and are not willing to play with others in a larger sandbox that could yield more public benefit. The question is how to link them with these larger public interests?

E. M. Forster, in his novel *Howards End*, said, "Only connect!" Indeed, it is this initial risk taking that builds association between meaning and memory, between patterns of behavior and acts of conscience that cause us to reexamine the history of place, as I see in this story of a narrative garden. John Dennis helped me to initiate such conversations, as did others whom I have acknowledged in these pages. His life bore witness to sites of conscience in Soweto near Johannesburg, and at Angkor Wat in Cambodia. I happened to run into him there, with his son Andrew, when looking for possible restoration sites with the World Monuments Fund in 1993.

I stopped at the National Trust for Historic Preservation annual meeting in Denver on my way to honor John Dennis in Corvallis. I wanted to meet the winner of the Crowninshield Award, which I had been nominated for by a now-deceased friend, Ruff Fant. At the ceremony I discovered the winner was Ruth J. Abrams, founder of the Tenement Museum in New York City; she spoke eloquently about creating an international center for sites of conscience, which my friend John Dennis had pursued in his own life.

Minutes before the awards ceremony in Denver, I was photographing the facade of the Brown Palace Hotel, only to be robbed in broad daylight while checking to see if the ornamentation correlated with the design of a newspaper-stand screen only a few yards away. (I had recommended such connections in my series of three sequential publications on this subject, *Place Makers*, begun in 1982.) An Asian woman pulled my phone out of my hands and ran across the crowded street. My heart pounding, I staggered after her in riding cape and cap, flailing my arms. Blind on one side, the effects of my last stroke, I dodged cars. My phone contained my instructions for arriving in Portland that evening and Corvallis the following morning, and I needed it. Confronted, but not blocked, by a potential witness on the steps leading to an office across the street, my would-be assailant suddenly stopped, turned around, and slowly pointed first to my still-trembling hands and then to a bollard where she placed the phone and said, "Here it is."

It was an act of redemptive grace, as I recalled next day in my remarks at the reception following the memorial service, which I felt was a matter of honor that I attend. I found a theme in this description of the actions of my erstwhile predator. I realized then that the making of a humanistic garden could itself be an act of redemptive grace—though I didn't have the presence of mind at the time to say that to the congregation.

The observations of the French aristocrat Alexis de Tocqueville in his work *Democracy in America* are famously penetrating. America, he said, will only become a great nation if self-interest is "rightly understood."[5] In other words, only if we see that it is in our private interest to serve a deeper public interest will we, as a society, rise above both ambition and self-serving greed, not anchored by long-term institutional memory—which De Tocqueville obviously thought an aristocracy provided.

That act of graceful redemption on a Denver street spoke to a larger and timely concern. My lost friends, acknowledged here, sought in their own eleemosynary actions to take the long view. This is also what we attempt to do with the humanistic interventions of narrative gardens anchored to place history and family memory.

Notes

1 Allen Gray Dunnington, *Forgotten as a Dream: Remembrances of a Virginia Lady* (Orange County, VA: publisher not identified, 1990).
2 Ronald Lee Fleming and Melissa Tapper Goldman, "Public Art for the Public," *The Public Interest*, no. 15 (Spring 2005): 55.
3 Martin Burgess Green, *The Problem of Boston: Some Readings in Cultural History* (New York: W. W. Norton, 1962).
4 Andrew Sorkin, "Ex-Corporate Lawyer's Idea: Rein in 'Sociopaths' in the Boardroom," Dealbook/Business & Policy, *New York Times*, July 29, 2019.
5 Alexis de Tocqueville, "How the Americans Combat Individualism by the Principle of Self-Interest Rightly Understood," in *Democracy in America* (London: Saunders & Otley, 1835–40), vol. 2, section 2, chap. 8.

6

Fig. 6 The "cautionary" steps in the children's area at Bellevue House spell out the qualities we hope to pass on to the next generations through the family narratives embedded in the garden

Foreword

This is the story of my narrative garden—actually, a series of twelve gardens. Narrative gardens speak to us about history and context, in a conversation that conjures up a landscape of place memory; they resonate with our past experiences and evoke memories of other garden places familiar to the seasoned garden visitor. The book traces my inspiration and motivation, exploring the gardens, rich with the cultural residue of recollection, that prompted me to eventually pursue my own garden dream. This residual set of place memories became rooted in one place—Bellevue House, in Newport, Rhode Island.

My quest was to go beyond the abstraction of color drift, floral vistas, parterres, and pathways to reimagine something that engages the mind as well as the eye in anchoring and honoring family memory over time. I wanted to use my garden to discover the cosmology of family life, to evoke the passage of time that resonates through it, to better understand and celebrate the rituals that can define and shape it. I sought to collaborate with artists and artisans who crafted both emotive content and architectural details that enrich the garden and its narrative, and build a commentary about it—sometimes bawdily humorous (as monkey sculptures can be), and sometimes ironic, I hope even subtle.

By discussing the techniques of reimagining this recollected landscape at Bellevue House, I hope to inspire others to create their own vision in their own landscape—to stimulate others to seek and find their own sources of placemaking and to calibrate their own sense of *locus*. At best, it should help others to break their own "mind-forg'd manacles"[1] and thus to lay down their own deeper claim to garden spaces.

Just as this book should serve as a guide for those establishing their own sense of family and place in the garden, so should it help others sharpen their capacity to see—to develop "the conscience of the eye," the means by which we inform our vision in a society that does not train us in visual literacy.[2] In framing an aesthetic composition, not only do we have to constantly edit out visual distraction to find that "perfect *National Geographic* view," but we also need to understand how to compose the integrity of that view. Before attempting to transform the built environment, we need training in how to make visual choices, and how to understand a visual language.

This book goes beyond the distilling of the sublime in a garden design—it attempts to better understand mistakes I made along the way in trying to achieve sublimity. A larger argument about compositional integrity as the best measure of authenticity is also addressed. Historic gardens and their buildings are better served by reasserting the integrity of context. Attempts to impose historicist revisionism, with its ideological assumption that there is a correct style for our time, has produced awkward juxtapositions in the renovations to Bellevue House. The garden, however, learns from contemporary experience. It avoids these clumsy juxtapositions of modern or postmodern design, and demonstrates that we are not bound to repeat the visual faux pas of the recent past through ideological conformity to an earlier and now discredited version of contemporary modernism. It argues that the authenticity of place is the predominant value.

Consequently, this book is different from a more traditional homage to contemporary gardens or a review of historic gardens. Of course, it draws on the work of those who have documented the great historic gardens in the past. Indeed, there are references to, and influences from, other gardens—from Little Sparta in Scotland to Studley Royal Park in Yorkshire; from the pebble shoal at

7

the Katsura Imperial Villa, Kyoto, to the stone water table at Villa Lante (figs. 7 and 8) at Bagnaia, near Viterbo, to the clipped topiary of Belgian landscape designer Jacques Wirtz and the architectural hedges that English designer David Nightingale Hicks carved out of yew and hornbeam.[3] Inspiration is also drawn from such tiny garden details as Achille Duchêne's mortised lintel joints at the National Museum of Decorative Arts in Buenos Aires, though we pegged instead of mortised our joints in our American Renaissance Water Garden. The carved monkeys shooting billiards on the lintel of the former Kildare Street Club in Dublin inspired the *Billiard Buffo* fountain in the formal French Garden designed by Achille Duchêne (figs. 9 and 10).

Fig. 7 Water table at Villa Lante, Bagnaia, Italy

Fig. 8 Villa Lante-style table at Bellevue House

8

9

Fig. 9 Monkeys playing billiards, stone carving by James and John O'Shea, Kildare Street Club, Dublin

Fig. 10 *Billiard Buffo* fountain, Bellevue House

At a time when most modern garden designers are interested in the abstractions of color, form, and texture, I find myself looking to the richness and complexity of those designers who learned from the Renaissance and kept its spirit in their work. Harold Peto, Edwin Lutyens, and Cecil Pinsent, superb English architects, recognized that there remains room for invention in older architectural languages and cherished the vernacular—Lutyens in particular. They even successfully mentored each other: Peto briefly hired Lutyens, who in turn employed Pinsent.[4]

Modernism has long been part of garden design, and an early modern garden designer such as Christopher Tunnard

10

would probably be astonished at the pervasive influence of modernism on garden design today, from kidney-shaped pools to slab benches and the insertion of abstract sculptural forms.[5] When looking at the portfolios of today's landscape designers or examining the issues of *Landscape Architecture Magazine*, one comes away dazzled by the slickness of it all. Certainly, there can be poetry in minimalism, but often the specter of minimalism appears to have reduced the design vocabulary to a sort of Esperanto. As critics of modernism have quipped, "Less is often a bore." Yet most of the gardens in magazines today continue to reveal an absence of historical reference and artisanal complexity; seldom do they express a connection to a historical example or focus on crafted details—an intricate drainage carving refracting off a larger pebble mosaic (figs. 11 and 12) or a well-wrought hinge (fig. 13), as found in the Bellevue House herb garden near the kitchen.

Largely missing from contemporary garden publications are gardens that seek to describe family life and empower family ritual in garden space—the sort of placemaking that is now increasingly called for in requests for proposals (RFPs) in urban design projects.[6] Yet private garden space has a less fractious or contested memory. This space gives opportunities to celebrate those elements of family meaning that can be refined and made subtle, assisted by the artist and artisan.

In pioneering numerous townscape projects in New England cities over a forty-year career, I came to appreciate the history and continuity of design elements, the rhythms and patterns of a design language. Too many modern gardens, however, show no love of craftsmanship and evidence of careful thought in their execution. When the narrative of a garden space is not linked throughout with

11

12

13

Fig. 11 Pebble mosaic in the Kitchen Garden

Fig. 12 Drain in the Kitchen Garden

Fig. 13 Wrought hinge in the Kitchen Garden

a continuity of design language, there is no resonance of refracted meaning.

The notion of gardens embedded with iconographic content stretches back into history. In Renaissance times, "family-proud gardens" celebrated the prowess and lineage of their creators, though often in clumsy, simplistic, even banal ways.[7] The display of weighty coats of arms and heraldic devices embedded in walls (which were sometimes crudely designed or constructed), or in clunky armorial garden topiary or furniture, characterized garden arrogance. The creation of grottoes—natural or artificial caverns housing sculptures—was an introduction of Mannerist style first to Italian and then to French gardens of the mid-sixteenth century (figs. 14 and 15). Even those that celebrated reigning families were often generic, as were some that shaped apertures in their ground.[8] Many are coarse assemblages of coral and volcanic rocks without particular craft or display of wit. Coats of arms and family mottoes do not of themselves make a great garden, nor do they reveal much of a story today for an audience of tourists.

There are, of course, some competitions between powerful families that generated an enriching rivalry of

14

15

Fig. 14 Diana of Ephesus fountain, Villa d'Este, Tivoli

Fig. 15 First room of the Buontalenti Grotto, Boboli Gardens, Florence, Italy

16

Fig. 16 Temple of British Worthies, Stowe House, Buckinghamshire, England

Fig. 17 Hellfire Caves, West Wycombe, Buckinghamshire, England

garden artifacts and follies that could sustain a narrative. Lord Cobham, a Whig of ambition and pretension with his Temple of British Worthies in the gardens at Stowe House in Buckinghamshire (fig. 16),[9] is a pious contrast to his agnostic neighbor Sir Francis Dashwood—less known for his restrained, classical gardens and more for his activity with his Hellfire Club of decadent humanists, engaging in anticlerical rites and rituals with country maidens (fig. 17).[10]

But such gardens are really exceptions, even in the English countryside, where the respect for eccentricity and differentiation of intimate garden space stands in marked contrast to the formulaic garden design found on the Continent. In France and Germany, there is often a radiating and manicured landscape of topiary parterre, pollarded tree columns, and recycled classical imagery that structures a power relationship between châteaux and schlosses and their manicured grounds without communicating very much about the humanity of differentiation.[11]

Eccentricity in gardens usually requires independent means and an independent vision. The folly at Dunmore Park in Stirlingshire, Scotland, crowned with a pineapple, is a striking example (fig. 18).[12] Generations of the Hoare family have spent their banking fortune improving Stourhead, an estate in the English county of Wiltshire; and John Aislabie, even after the failure of the South Seas

17

Company, retreated to North Yorkshire to continue work on Studley Royal Park, where temple follies line limpid waterways.[13]

This drive to achieve a certain excellence no matter the cost, which I so admire, expressed itself again in the extravagant nineteenth-century collaboration between the 3rd Marquess of Bute and his architect William Burges, who remodeled Cardiff Castle in Wales to express their admiration for the Middle Ages. The rooms celebrate their love for exoticism and color with complex astrological symbolism. There was an Arab room, a Roman court with a roof garden, and a summer smoking room, which indulged their interest in religion, mysticism, and the unseen. While the gardens on the castle grounds didn't share these exceptional qualities, the castle itself, employing scores of artists and artisans, became essentially the world's largest folly, all to suit the marquess, described as the richest man in Europe.[14]

Another example of British eccentricity was Edith Vane-Tempest-Stewart, Marchioness of Londonderry, whose Dodo Terrace in the gardens at Stewart House in County Down, Northern Ireland, features cast-concrete sculptures depicting the animal names she gave her Ark Club friends, aristocrats who were personally invited to her supper entertainments during and after World War I; Winston Churchill was cast as a walrus. Today Michael and Anne Heseltine continue the tradition of turning from the world of politics and commerce to create their Thenford Gardens and Arboretum in Northamptonshire, including their extended rill and mosaic-lined pool, much larger than those found at Bellevue House.[15]

My garden is an American garden, reflecting the experience of an American family. It is not the heavy tread of European armorial symbolism marching through an entry hall, but the more modest intersection of the artifacts of American experience: the cloth cap of an ancestor killed at Chancellorsville, the image of a buckboard in the Oklahoma Land Rush, the orange crate designs and the smudge pot recalling the citrus groves of Southern California, the aspirations of a child with books about ghost towns who badgered his parents to drive dusty western roads to discover them.

In designing our own garden, my family and I found an opportunity to tell richer stories about place, with a deliberate plotting of ritual that can consolidate memories around it, carefully supported by an armature of placemaking. We discovered that working through these ideas gave us opportunities to integrate meanings that transcended physical planting patterns and the arrangement of flowers and trees. We aspired to tell stories that demonstrate what had shaped our very desire to create a garden realm.

I grew up in an age of urban renewal with its concomitant loss of identity, which was to haunt the American cityscape. I sought in my professional work to redress this sense of loss and so devote myself both to placemaking projects and to understanding how feelings of proprietorship could redress these losses, how a sense of ownership could build a sense of accountability in place. My previous books on placemaking in public spaces (*On Common Ground* and *Place Makers*) and on building facades (*Facade Stories*) sought to create an emotive link between people and place.[16] What could we do to strengthen that bond? The old greens and commons had proprietors. In *On Common Ground*, Lauri Halderman and I identified through case studies how that attitude of proprietorship could be fortified in public spaces today. *Facade Stories* showed how people could pay attention to the changing role of building facades. In these three placemaking books, we demonstrated how objects could both enrich the story of a place and amplify our affection for it. All of these approaches could be applied to a private garden, which becomes increasingly a public space to enjoy, as we discuss in chapter 3.

My family's private garden is enriched by integrating the work of artists and artisans. A series of placemaking elements cast in bronze tells some of our stories: a cultural

Fig. 18 The Dunmore Pineapple, Dunmore Park, near Airth, Stirlingshire, Scotland

18

landscape of landmarks from Southern California that our parents, grandparents, and great-grandparents all knew; a series of vignettes symbolizing ancestors on the sides of a pizza oven chimney. These elements are part of a larger narrative that documents what has been lost.

I also wanted to explore the peculiar motivating force that secured my interest in place from an early age: my childhood love of ghost towns, antique weapons, and artifacts from my grandparents' trip around the world in the 1920s. I commissioned some thirty-five artists and artisans to celebrate aspects of garden making and explore some larger questions: Why did our culture seem to ignore the integration of art and architecture, not just in gardens, but throughout the built environment? Why did other cultures appear more vibrant? Why is there not a more comprehensive integration of craft and modest and utilitarian public art in most public architecture?

The idea for a strategy of placemaking in the garden came to us incrementally. We wanted our family to discover its own history, and, having a fragmented family, we looked for opportunities to invent new rituals that could connect our family back together. We also wanted to enrich and strengthen our bonding through an engagement with the arts, particularly with dance. To refashion a phrase, we wanted to "dance to the memory of lost time" and to refashion the gardens to support new meaning with new rituals.[17]

The site of this adventure is an early-twentieth-century house and gardens on Bellevue Avenue in Newport, Rhode Island. The Federal-revival house stands on grounds that were partially configured by the French landscape architect Achille Duchêne. Aside from very large specimen trees, a two-story McIntire-style tea house, and a gazebo built into the garden wall, it was an open canvas. Only gradually did we come to understand what to do about it, and therein lies our tale.

Chapter 1 recounts our first encounters with the house and grounds. Like travelers from afar, we needed to find our bearings in the garden, to walk the metes and bounds and locate the principal sites and monuments. We needed

19

Fig. 19 Violin performance at Bellevue House. Siena and Mathieu de Menonville can be seen in the foreground.

to discover what framed the garden and what tied it together in order to connect and refine the garden spaces. In architectural terms, we call this vocabulary of elements the *pattern language*. Gordon Cullen and Ian Nairn speak about this collection of parts in their books on townscapes.[18] A memorable place is not just an arrangement of landmark vistas but a *tout ensemble* with integrated, harmonious transitions from viewshed to vista. It was our own effort to define those qualities that produced a Calvino view.[19]

Chapter 2 then describes the search for family memory, its poignance and its elusiveness, and the irony that sometimes emerges in efforts to recover it through didactic exercises like cultural landscapes and "family chimney trees." The garden also evokes memory through the quoting of famous gardens from around the world, building a design vocabulary that has a capacity to resonate as it brings to mind images of other places. These images are sometimes subconscious, so the juxtapositions may help the casual garden goer to discover them for the first time. The gardens can thus speak to us by projecting metaphor and allusion and even occasionally by talking back.

This "talking back" includes expressing the spirit of animation. Chapter 3 explores how programing a garden can evoke its spirit (fig. 19). For the most part, this is a momentary phenomenon that can be both vivid and transitory: a dance around the limbs of an ancient tree; the choreographing of a fountain to express the water sprites that might occasionally inhabit the gardens; or a sound portrait, such as that deployed by the Italian composer and rosarian Walter Branchi to describe the garden with Rothko-like intensity through a tonal density of chords.

Chapter 4 explores how meaning and memory come together in the "academical village" in the garden, a site at which the physical structures become part of a larger mental landscape, populated with lost friends and war memories.

Chapter 5 discusses the place of art and craft in contemporary gardens, in both private and public settings. Fine details—hinges, bronze castings, even the design of a paving joint—can create a level of complexity, finish, and intimacy, and commissioned artworks can add to a personal narrative.

The final chapter, chapter 6, is about humor, both broad laughter and subtle irony. Humor—sometimes didactic, sometimes veiled—can be a challenge to achieve. A shift to miniature scale in a garden, for example, can be both amusing and sometimes disorienting; hidden grotesques can stir surprise and wonder, but they can also poke fun at the limits of our achievement over time or at the somber garden visitor. Through such devices can a garden finally talk back!

Notes

1 William Blake, "London," in *Songs of Experience* (1794).

2 Richard Sennet, *The Conscience of the Eye: The Design and Social Life of Cities* (London: Faber & Faber, 1991).

3 Nick Meers and David Wheeler, *Panoramas of English Gardens* (London: Weidenfeld & Nicolson, 1991), 142–49.

4 Robin Whalley, *The Great Edwardian Gardens of Harold Peto: From the Archives of Country Life* (London: Aurum, 2007), 7.

5 Christopher Tunnard, *Gardens in the Modern Landscape. A Facsimile of the Revised 1948 Edition*, with a new foreword by John Dixon Hunt (Philadelphia: University of Pennsylvania Press, 2014).

6 Ronald Lee Fleming, *The Art of Placemaking: Interpreting Community through Public Art and Urban Design* (London: Merrell, 2007).

7 Indeed, there were predictable places to find these linkages; even today we can find them in a modest scale in English country houses. One can usually discover the provenance of old public-school boys in their house pictures, often located in the loo behind the front hall stairs. There we see Harrovians, Old Etonians, and Wykehamists immaculately uniformed with their house masters enforcing gravitas.

8 Christopher Thacker, *The History of Gardens* (Berkeley and Los Angeles: CA Press, 1979).

9 Michael Bevington, ed., *Stowe: The People and the Place* (Swindon: The National Trust; London: Scalia Publishers, 2011).

10 Jason M. Kelly, *The Society of Dilettanti: Archeology and Identity in the British Enlightenment* (New Haven and London: Yale University Press, 2009), 74–76.

11 Russell Page made a similar observation on the difference between continental and old English gardens; see Russell Page with Fred Whitsey and Marina Schinz, *The Education of a Gardener*, American ed. (New York: Random House, 1983), 95.

12 Gwyn Headley and Wim Meulenkamp, *Follies, Grottoes and Garden Buildings*, rev. ed. (London: Aurum, 1999).

13 Studley Royal Park was, in fact, the model for the formal English side of the miniature Oriental Vale at Bellevue House.

14 Matthew Williams, *Cardiff Castle* (London: Scala Arts, 2008).

15 Anne Heseltine and Michael Heseltine, *Thenford: The Creation of an English Garden* (London: Head of Zeus, 2016).

16 Ronald Lee Fleming and Lauri A. Halderman, *On Common Ground: Caring for Shared Land from Town Common to Urban Park* (Harvard, MA: Harvard Common Press; Cambridge, MA: Townscape Institute, 1982); Ronald Lee Fleming and Renata von Tscharner, *Place Makers: Creating Public Art That Tells You Where You Are*, 2nd ed. (Boston: Harcourt Brace Jovanovich, 1987); Ronald Lee Fleming, *Facade Stories: Changing Faces of Main Street Storefronts and How To Care for Them* (Cambridge, MA: Townscape Institute; New York: Hastings House, 1982).

17 Anthony Powell, *A Dance to the Music of Time: First Movement* (Chicago: University of Chicago Press, 1995); Marcel Proust, *Remembrance of Things Past*, trans. C. K. Scott Moncrieff and Terence Kilmartin, 3 vols. (London: Vintage Books, 1982).

18 Ian Nairn, *The American Landscape: A Critical View* (New York: Random House, 1965); Gordon Cullen, *Townscape* (New York: The Architectural Press, 1961). Thomas Gordon Cullen (1914–1994) was an influential English architect and urban designer who was a key motivator in the townscape movement. He is best known for his book *Townscape*, first published in 1961, writing as Gordon Cullen; the book was published in the United States by Van Nostrand Reinhold in New York in 1962. Later editions were published under the title *The Concise Townscape*—published again by the Architectural Press, as an imprint of Routledge.

19 Italo Calvino, *Six Memos for the Next Millennium* (New York: Mariner Books, 2016), first published 1988. Calvino's key qualities are lightness, quickness, exactitude, visibility, multiplicity, and consistency.

Chapter One

NAVIGATING THE COSMOS OF FAMILY

Piecing together the landmarks that define it and the patterns that connect it

The idea that my family and I would own a secret, walled garden in the middle of a city would have never occurred to me a few short years ago. I was neither plantsman nor garden designer—mowing a small plot of grass in Cambridge, Massachusetts, where we lived, was about as close as I came to horticulture. My then wife and I were interested in townscape design and writing about community character; we had three young children to educate. We spent as much time as we could taking them on European trips, including travel to Switzerland where my former wife's family owned a *Stökli,* a small vineyard house, and a very small, very ruined castlette on a hillside in Basel-Landschaft.

When my wife and I divorced, I, with the guidance of a charming and wise older woman, who was a long-time resident of Newport, found comfort by the Rhode Island shore. I was introduced to a new community and finally advised to look at Bellevue House, a Federal-revival mansion—a "cottage," as Newport mansions built in the Gilded Age era were described. And so, in a low market, and in a low state of mind, just two weeks after my domestic difficulties were legally resolved, I made a bid on Bellevue House (fig. 20).

Children's fountain in the American Renaissance Water Garden, Bellevue House (detail, fig. 100)

The house headed the parade of "cottages" that lined Bellevue Avenue where elegant carriages and liveried coachman engaged in a daily ritual of seeing and being seen. It stood at the northern end, next to a shopping center, and had been empty for a few years; developers were eyeing it speculatively as a possible site for a boutique hotel. It was a beauty of a house, but maintenance had been much deferred. The garden was full of great specimen trees and the plowed-under flower beds were a barely perceived outline in a vast sea of grass that needed to be cut every week. Who would have known that the three and a half acres behind the wall was a private world, a hidden oasis? It was a large garden even by Newport standards. My children and I were told there had once been three gardeners; now there was a crew that came once a week to cut that grass and trim the hedges, an owner, and a caretaker, who, we later discovered, concealed five inches of dust under his bed in the staff quarters. I soon discovered that before we could seriously think about restoring, much less reimagining, the garden, we had to think about repairing the house.

The designer of Bellevue House, architect Ogden Codman Jr. (fig. 21), was a co-author, with writer and designer Edith Wharton, of one of the first books on interior design in America, *The Decoration of Houses* (first published in 1897 and still in print today). Codman

20

21

22

23

Fig. 20 Bellevue House, circa 2018. Only the large house at the top of the photo and the two-story McIntire teahouse at center right were here in 1999 when I bought the house.
Fig. 21 Ogden Codman Jr., 1860s

Fig. 22 Martha Codman, portrait by Georgine Campbell, 1906, of the Museum of Fine Arts, Boston, Bequest of Maxim Karolik, 64.617
Fig. 23 Jane Pickens, former owner of Bellevue House, 1960s

designed the house in 1910 for his third cousin, Martha Codman (fig. 22); singer Jane Pickens would later own it (fig. 23). The dimensions were based on three Georgian- and Federal-style mansions in Roxbury, Massachusetts, which Codman had measured when he was a student at MIT.[1] He finished the drawings, but never the course. He apprenticed with his uncle, architect Robert Peabody of Peabody & Stearns.

The severity of the house's Federalist exterior did not prepare us for the glorious rotunda at the central axis of the interior, its cantilevered staircase sweeping upwards to the Adamesque dome forty-four feet above the elaborate inlaid marble floor (figs. 24 and 25). As we contemplated this soaring construction, however, we noticed fragments of plaster had fallen from the frieze at the apex of the dome. It appeared that seagulls had dropped shells on the glassed center light, cracking it. Water was gradually demolishing the delicate frieze work. My first great expenditure would be to build scaffolding in the center of the rotunda so that the skylight could be replaced and the surrounding plasterwork refashioned and repaired. There was no time for contemplating garden design—to neglect the ceiling was to lose the delicate frieze.

24

Then there were mundane house problems that further distracted me from the garden. Codman designed Bellevue House as a summer house, but we saw it as a year-round

25

Fig. 24 Renovated rotunda, Bellevue House

Fig. 25 Staircase

weekend retreat, so it had to be insulated against the New England winters. As in all large houses of this period, the cook occupied the pantry and kitchen, and the mistress of the house, Miss Martha Codman, as she was then, would only appear briefly in the kitchen to work out the menu. Thus overlooked, the kitchen retained its unholy trilogy of vinyl, plywood, and linoleum finishes dating from the last renovation in the 1960s or before (figs. 26–28). We replaced the plywood cupboards, inserted pilasters, installed wainscoting, and added columns of similar proportions to those found in the public rooms.

At the age of sixty-three, Martha Codman married a thirty-three-year-old Russian-born opera singer, tenor Maxim Karolik, and together they went on to assemble the first great collection of eighteenth-century American furniture and nineteenth-century paintings and drawings. They subsequently donated their collections to the Museum of Fine Arts in Boston, where they became the foundation for the American Wing. Poor Ogden Codman Jr., for other reasons, went off to self-exile in France. Pauline Metcalf, editor of *Ogden Codman and the Decoration of Houses*, reported that the children of his wife, a wealthy widow, told her that Codman had hijacked their trust funds to build Villa Leopolda, reportedly the most expensive house ever constructed on the French Riveria.

When my three children, all excellent cooks, scrutinized the kitchen that the Karoliks left behind, they were not amused. They saw the kitchen and the adjacent staff dining room (a potential media room) as unacceptable, and they informed me that the kitchen would have to become the new center of the house. They wanted a family center and, like most contemporary American families, they saw that center in an extended kitchen. The elegantly proportioned dining room and living room would remain for large dinner parties, but the family would be hanging out in the kitchen. And so it came to pass that another two years went by while J. P. Couture, the Providence architect, completely redesigned and reconfigured this space. We were still not considering the gardens.

New plastered and columned door frames, which extended the symmetry of the carefully detailed porticos of the public rooms, were fitted around a large green-and-pink-veined marble countertop, which the children could gather around (figs. 29 and 30). We achieved a first reference to a future garden in a new tile splash panel over the stove. It was a scene of the garden to be, complete with a cupola bracketed by Federal-style structures, and depicted the children at play (fig. 31). Within a few years we would complete that cupola, as described later in this story.

Figs 26–28 The old kitchen, with 1960s fixtures and finishes

26

27

28

29

30

Figs. 29 and 30 The new kitchen, with Reynolds Fleming

Fig. 31 Splashback over the new kitchen sink by James Barnes and Deborah Scott-Wilson. The tiles reveal an imagined garden vista braced by monkeys supporting a coat of arms.

31

32

One part of the renovation brought us closer to a vision of the garden: the reinforcement of a pantry that became a breakfast room and opened eastwards toward what could become a garden (figs. 32–35). We adopted an original Codman design that he had proposed, but never implemented, for the plasterwork of the ceiling at Edith Wharton's Land's End in Newport. Adding a skylight and echoing the columns in the dining room gave this little room a special elegance. Importantly, it provided a long vista looking eastward into the old Yew Garden, with its charming little gazebo dating from the origins of the house and built into the northern wall of the property.

Finally, we were able to make a break for the green with the creation of a new outdoor room that would serve as an adjunct dining space on the eastern side of the house, where there had been only nondescript lawn, a cement wall (originally brick, as an old photograph reveals) separating the Yew Garden space, and the clothes line adjacent to the house. Just outside the French doors of the new breakfast room, formerly a pantry and coal shoot, we constructed a paved stone terrace (figs. 36 and 37). This was to be a social space, with a sturdy stone table, and an elevated vantage point from which to contemplate the vista that was beginning to emerge, better enabling us to view the corridor that drew the eye back to the gazebo.

Having achieved a vantage point of steps and terrace, we next defined the vista to the east with two new stone-capped brick columns identical in design to others demarking the walls to the south. This linear space flanking the east side of the house would become our American Renaissance Water Garden.

Figs. 32–35 Breakfast room with Louis XVI chairs purchased from a yard sale at another Newport mansion, Bois Doré, and cove ceiling originally designed by Ogden Codman Jr. for Edith Wharton

Figs. 36 and 37 New garden view from the breakfast room

33

34

35

36

37

Defining the parameters of the garden

Before beginning to construct or renovate a garden, the wholeness of the garden composition must be examined, its form seen in the mind's eye. The pattern language that holds it together must be identified, then used to build rhythm, generating drumbeats of repeated forms to keep the eye moving forward. In the Bellevue House garden, this means that the eye connects latticework with fence lines, clipped hedges with tree corridors. Enclosed vistas frame the visual splay and splash, and the aural gurgle and murmur, of fountains and pools, shaping the tempo of the garden: we hear it as well as see it. Architectural elements—volutes, mosaics, and even carved drain covers—are repeated to build complexity, intricacy, and richness. These parameters of place all help to define the cosmos of the garden. Once this initial vision is consolidated, we can introduce the call and response of memory, the spirit of animation, art and craft, and finally humor and even irony.

Defining the identity and locational boundaries of Bellevue House advanced our understanding of place. In this particular garden, memory of the historical context within this location is the predominant value we celebrate. The house pays homage to Samuel McIntire, the wood carver who became a self-taught architect. He designed some of the most graceful buildings in the Federal style for the great merchant princes of Salem, Massachusetts—the Junta, as they called themselves. The Junta included the leading figure Elias Hasket Derby, the richest of these merchants. It was for Derby's great-great-granddaughter, Martha Codman, later Karolik, that Ogden Codman Jr. designed Bellevue House, or Berkeley Villa as it was first known.[2]

Derby achieved extraordinary renown. His privateer opened the Cape of Good Hope to trade in 1784, just after the American Revolution. The following year, he made the round trip to China, and he was also the first to reach Bombay. Salem's fortunes blossomed with the subsequent trade in spices, ivory, silk, and jade, and the town was able to build fine houses on Market Square. Derby hired the carver Samuel McIntire, who had designed several great houses there.[3]

This was a period of worldwide focus on classicism, which arose from archeological discoveries in Greece and Rome. In New England it was less the imitation of classical correctness, as seen later in the Greek Revival, but rather an effort to imitate classical forms and qualities. Walls were made of stuccoed brick, and masses became more geometric and symmetrical with the addition of balustrades, parapets, and bold, projecting curved bays. The addition of round forms brought in a new grace of movement. The elliptical fanlight became a familiar overdoor motif that surmounted balancing narrow sidelights. The projecting portico became another popular addition with many variations of slender columns, which the brothers Adam had introduced in England.

The *sine qua non* of the interior was the ascending spiral staircase enclosed by oval-shaped rooms, often seen with projecting bays. Decorative motifs on doorways, mantelpieces, and ceiling cornices became more delicate and finer in detail than had been seen in Georgian architecture. These motifs were almost always painted in pure white, in keeping with the new spirit of chasteness. In the United States, the grace and refinement of the Federal period, seen in the work of McIntire and Charles Bulfinch, achieved a greatness of style that rivaled the Adamesque splendor of England. The Federal style ushered in a new lightness and unity of composition with restraint of architectural detail, a new classicism of elegance and assured mastery.

Building compositional integrity and architectural authenticity

Our reconstruction work on Bellevue House and the building of new structures in the garden brought us into contact with larger issues of architectural preservation and the fallout from the Venice Charter. The *International Charter for the Conservation and Restoration of Monuments and Sites* was drawn up in Venice in 1964 by a coalition

38

of international curators, archeologists, and modern architects seeking to develop guidelines for the treatment of historic buildings.[4] The group's objective was to safeguard monuments not only as works of art but as historic evidence.[5] The charter stressed the differentiation and preservation of each stage of development, requiring that any necessary restorative work "must be distinct from the architectural composition and must bear a contemporary stamp."[6] This effort to both protect historic context and avoid false history attempts to give equal weight to these sometimes contradictory standards and takes a narrow view of time and authenticity, honoring a rigid historicism in which each period has its own architectural style to be saved without critical evaluation of its impact on the legibility of a larger culture landscape.

39

Fig. 38 Warsaw old town, Poland, 2012

Fig. 39 Amoeba-shaped Graz Art Museum, Austria, in the middle of a historic setting

The charter implied that a forty-year span of high modernism superseded earlier architectural styles.[7] It did not comprehend the living tradition of architecture, which, like the Taos pueblo, demonstrated dynamic and organic growth. The massive restoration of monuments in communist Eastern Europe initiated before the Venice Charter, such as Warsaw's old town square during the Second World War, sought to recreate historic settings damaged or destroyed, allowing citizens to retain their sense of place identity (fig. 38). In our judgment, the desire of Warsaw citizens to preserve their identity argued for a higher definition of authenticity, one that honored historic forms rather than progressive change. Dates can always be inserted into a restoration project, but the higher value is preserving the investment in the essential integrity of building form. The charter committee, however, would have new stones differentiated from the old in a reconstructed building; they would put new blank clay in a reconstructed Greek pot from an archeological dig. It was sometimes thought that freezing preservation and assuming modernism was the appropriate architectural expression of this age. The Venice Charter, with its intention not to create "false history," was itself frozen in time.

In our own townscape work we had noticed the disconnect between those who would preserve old buildings and those who were practicing modern architecture. As streetscape planners working in small cities in New England, we were aware of the most extreme modernist invasions of traditional cityscapes—"the porcelain exclamation mark of a gas station at the end of a Federalist sentence," as we put it to citizen groups concerned with the enhancement of Portsmouth, New Hampshire.

We saw that many preservation professionals were committed to saving old buildings but were not concerned with systematically examining how old architecture worked to create a design pattern language. Some accepted discontinuity as an example of historicist change, fostered by modernism, and made no value judgment about the impact of the disruption. Laymen were more interested in protecting the essential character of the city. Apparently, Portsmouth had more eighteenth-century buildings than Colonial Williamsburg and could have been a candidate for holistic preservation. It was this historicist notion of accepting the juxtaposition of old and new that seemed to widen the fatal gap between preservation and modern architecture.

An extreme example is the intrusion of modern-style gas stations in historic settings. We produced a book for the American Planning Association entitled *Saving Face: How Corporate Franchise Design Can Respect Community Identity*, which showed examples of compatible design where fast food enterprises and gas stations respected the existing cityscape, often as the result of design review laws that supported this change in consciousness.[8] Instead of a few carefully curated monuments and a small number of nationally significant districts, there was now a widespread movement toward compatible design in which the modern sits comfortably alongside the historic, which in turn has to be seen as a part of history as well! So we found ourselves advocating for a contemporary view of historic progression. We discovered we could learn from the dogmatic approach of the Venice Charter to preserving all evidence of change.

The *Secretary of the Interior's Standards for Rehabilitation* (1977) followed the language of the Venice Charter.[9] But because they were being primarily used to evaluate buildings applying for tax credits, these standards often demonstrated more flexibility to accommodate adaptive use. Indeed, the Back to the City movement involved vast numbers of laymen who wanted to save the fabric of traditional neighborhoods and took issue with the narrow approach of many preservationists and the utopian strategies of planners and modern architects. Both architects and preservationists failed to recognize the traditional forces that shaped and supported compatible design solutions and did not respect time-honored ways of building.

Yet the times have changed, and the standards need to change to accommodate new generations who respect the craft of old places. A whole new industry has grown up to accommodate their desire to use traditional building materials and decorative elements, from towel racks to street furniture. There is now a growing national movement of designers formed in chapters of the Institute of Classical Architecture and Art who are militating for change.

Meanwhile, only two architectural schools, Notre Dame and the University of Miami, are training students to work with the full range of architectural styles and promote the idea that this is a vibrant language that all designers should understand. Most schools remain bastions of modernism, and modern designers still command influential commissions and promulgate their modern style that continues the disruptive doctrine of contrast and conflict. Brent Brolin, whose brilliant book *Architecture in Context* (1980) demonstrated how modern structures could be redesigned to fit into a cityscape, was a pioneer for a changing sensibility.[10]

But now there are several generations of trained preservationists who maintain the disruptive ideology of modernism as an act of faith, even ideology. Robert A. M. Stern's new colleges at Yale, which use a traditional Georgian design language, are the most outstanding example of a shift in sensibility. These buildings could not have been commissioned in the early 1960s when Eero Saarinen designed Yale's Ezra Stiles and Morse Colleges, abstracted stone forms supposedly related to the earlier Gothic.

The economic pressure to develop Portsmouth finally resulted in the demolition of the gas station, and the construction of a new brick structure that allows some healing grace. But there are still rigidly historicist preservationists out there who would save the telephone poles clogging the traditional sight lines of Civil War battlefields, struggling to save a contextually inappropriate Richard Neutra-designed visitor center at Gettysburg that the National Park Service itself finds no current use for and has been seeking to demolish. And there is still the conundrum of preservationists protecting a period of design that did not respect its context, while a more environmentally conscious generation wants to protect the context itself. This battle is still being fought, but it is now aided by the concept of large conservation zones, introduced with the Malraux Law of 1962 in France and the Civic Amenities Act of 1967 in Great Britain—a private members' bill led by Lord Duncan Sandys, Churchill's son-in-law—whose broader definition of conservation areas encompassed whole villages and towns.

The earlier contrasting and conflicting interpretations of successive passages of architectural time and the preservation of distinct stylish differentiation that followed on from the Venice Charter would eventually set up historic districts as sitting ducks for dramatic interventions. Modern architects can disfigure the very historical assets that more recent, broader conservation legislation was designed to protect (fig. 39).

During the renovation work at Bellevue House, because of time constraints, we had to follow the local Preservation Commission mandate, and make a modern "stub" chimney for our breakfast room addition (figs. 44–47). However, we firmly believe that historic gardens and their buildings are better served by reasserting the integrity of context. In the garden, authenticity of place is the predominant value. Bellevue House and its grounds extend the memory of Derby, the merchant prince. The new outbuildings repeat the Federal style of the original house, creating a historical framework of architectural and garden memory. The new architecture also shaped a space in the grounds for public use, creating a common ground that increased the animation of the property by expanding its function. Instead of creating a juxtaposition of different styles in the garden structures, we sought to extend a sense of harmony that sustained a rhythm—or even amplified that rhythm.

40

Creating a design strategy and establishing a pattern language

The design strategy of the garden begins at the front gate. The pattern language is established in the carefully worked pierced lattice panels of the gate, which became the basis for a repeating design motif. Codman appropriated the design for the gates at Berkeley Villa (as Bellevue House was first known) from the gate of Heyward House, which had been demolished in Charleston, South Carolina, soon after Berkeley Villa was constructed. Indeed, part of Codman's genius was to carefully utilize design ideas from the rich trove of architecture he had seen in both the United States and Europe.[11]

The design motif is based on a sketch Codman made of the garden of the Dutch palace Het Loo.[12] He intended to bring this motif into the original garden as part of the trellis design, but never implemented it. The pattern is repeated in the enclosure on the terraces behind the Villa Lante-style table adjacent to the house and on the terrace behind the emerald lap pool at the eastern end of the property, and echoed in the gates in the north and south garden walls, and in framing elements such as a small exedra that Codman initially created for Edith Wharton at Land's End (she later moved it to the topiary garden at The Mount in Lenox, Massachusetts) (figs. 40 and 41).

While some contemporary architects curl their lips and call this "pastiche," Codman recognized that a coherent architecture could be assembled from a variety of historical sources, as did English architect Edwin Lutyens in his buildings for English landowners and for the British Raj in New Delhi.

This idea of repetition and association is central to understanding the nature of the Bellevue House garden. The result is a far richer pattern language, delicate and quite subtle, that defines different garden spaces. While other gardens seek their greatest strength in arresting planting patterns, here the deliberate use of architectural elements—not simply as nodal points, but in a denser, more reiterative manner—adds a complexity and rhythm that many recent gardens lack.

Framing a common ground

When we began to bring structures into the garden—the McIntire teahouse from the Derby mansion in Salem, the cabanas and buildings of the eastern end of the garden—we found ourselves defining and enclosing garden spaces in a way we had not anticipated when we began the restoration and renewal of Bellevue House (figs. 42 and 43). As we gradually shaped the northern end of the garden with a greenhouse, a stable block (housing a conference room and guest bedrooms), and a library/nymphaeum, we discovered

41

that the new buildings had become more than the sum of their parts and that we had, in fact, created a sort of common—a space that we could open to the community.

This brought a new metaphoric overlay to the garden and gave rise to new questions about the nature of the garden and the spaces we were creating: Did this common space have increased social meaning? Could we view this common as analogous to others in New England? Was there now a social contract to extend the use of the space for public benefit? Seen metaphorically, the social contract is perhaps the most elusive problem that transfixes our larger society and presents a transcendent challenge for our commonwealth. Could increased physical harmony within a public garden space encourage more social agreement in the world beyond?

The common ground at Bellevue House reflects the harmonious pattern language of the white clapboard buildings in the Federal style. But shouldn't a perfect union of form and function transcend architecture? The common ground opens out from the eastern side of a formal allée into a viewshed of the academical village—a collection of buildings arranged around an open lawn, such as Thomas

Fig. 40 Trellis at Het Loo Palace, Apeldoorn, Netherlands

Fig. 41 Trellis and Pomona statue in the American Renaissance Water Garden, Bellevue House

Fig. 42 Bellevue House, c. 1922. The original brick wall (at bottom) had been replaced with concrete. We added columns, rebuilt a low brick wall, and planted a herbaceous border that separates the lawn rill from the Villa Lante-style table on the terrace. The rill terminates at the children's fountain at the south end of the American Renaissance Water Garden.

Fig. 43 Bellevue House, after 2017. Hydrangeas are blooming in the Yew Garden, and a brick wall separates the herbaceous border in the American Renaissance Water Garden.

42

43

44

45

Fig. 44 Original lattice work seen from the herb garden

Fig. 45 View from the herb garden, looking back at the extended, renovated kitchen

Fig. 46 Original chimney pattern and height respects the harmony of the composition and historical precedent

Fig. 47 New chimney stump contrasts and conflicts, responding to thirty years of high modernism rather than creating harmony as part of a whole

46

47

48

49

50

Jefferson planned at the University of Virginia. While Jefferson favored classical architecture, at Bellevue the structures recall the Federal style of McIntire's work (figs. 48 and 49).

The space is anchored by the high-style two-story teahouse, flanked by a white picket fence, gate posts, a pinnacled greenhouse, and a stable block with a loggia that attaches it to the new library/nymphaeum. The attention is then drawn to the south, where visitors first hear, then perceive the glint of a cascade, the Palladian form of a bridge, and delicate bathhouses (cabanas) separated by a little terrace space. Striped green and white awning spans the Family Chimney Tree, which stands atop a pizza oven wrapped in bronze foliage on which is recorded elements of Fleming and Reynolds family history. From here, the eye travels to an emerald pool below the terrace, which confronts a majestic rotunda framing the view from the library. This is an intricate vista of compatibly scaled Federal architecture. Like a New England village common, it represents a compact with proprietors to be seen and admired by the family, and to be enjoyed by the community.

51

This new common is framed by two monkeys contemplating each other with telescopes. One is fashioned in copper, perched as a weather vane on the reconstructed cupola of a long-vanished McIntire-designed church; the other is carved in wood from the great beech tree that stood adjacent to the house on its southern exposure (figs. 50 and 51). The wooden carving represents the philosopher Jean-Jacques Rousseau depicted as a top-hatted monkey with *The Social Contract* under his foot; he peers across the new common at the crafted copper sculpture swinging in the wind. The two figures pose the question: Can we find a common ground in an increasingly diverse and contentious society? The answer, according to Bob Dylan, is "blowin' in the wind."[13]

Fig. 48 Aerial view of Bellevue House from the east

Fig. 49 Aerial view of Bellevue House toward the northeast

Fig. 50 Monkey sculpted in bronze on the cupola

Fig. 51 Monkey carved from a beech tree

52

53

54

Repetition of architectural details

We started our quest for a larger coherence in the garden with only one central structure—the two-story McIntire-designed teahouse, which architect Fiske Kimball, dean of the University of Virginia and director of the Philadelphia Museum of Art, replicated for Martha Codman in 1922.[14] The original teahouse, built in 1793–94, was named the Derby Summer House; it featured a portico opening through the ground floor, and mounted on its roof were wooden sculptures (*acroteria*) of a yeoman and Pomona, the ancient Roman goddess of abundance.

Our structure, without Pomona atop, anchored our garden and provided the impetus for other McIntire-style buildings, repeating architectural forms that give the garden its basic pattern language and its rhythm. The small buildings are used to command focal points and viewsheds (figs. 52–58). The garden now features two teahouses, a cupola (replicated from the single surviving drawing of a burned church in Salem; see fig. 93), two Federal-style bathhouses, a library/nymphaeum, and the classically detailed Derby Coach House (conference room and guest bedrooms), as well as a pergola, gazebo, exedra, and an arch replicating (in smaller scale) the four McIntire constructed in Washington Square, Salem, as part of a city green enhancement project directed by Elias Hasket Derby.

In modern design, architectural elements are often used idiosyncratically to strike the occasional dramatic chord—sometimes, in fact, a dissonant note of asymmetry. But the effort of a pattern language is to build a seamless flow through space that carries the eye forward with richly evoked details punctuating spatial forms (figs. 59–66). At its best, a pattern language creates a harmony of accretion and of accent while developing and reimagining each new garden space.[15]

Figs. 52 and 53 Central allée, looking north

Fig. 54 Looking toward the formal French Garden and exedra

55

56

57

Fig. 55 Exterior of the allée

Fig. 56 Lovers' seat enclosure

Fig. 57 North side of the enclosure, looking south to the newly constructed small teahouse

Fig. 58 South lawn

58

59

60

61

Fig. 59 Side gate

Fig. 60 Front gate

Fig. 61 Fence embracing the old teahouse

Fig. 62 Original gazebo built into the north wall in the Yew Garden

62

63

64

65

66

Fig. 63 Kitchen Garden gate

Fig. 64 Lattice in the American Renaissance Water Garden

Fig. 65 New Codman exedra in the French Garden

Fig. 66 Central allée of the pergola, looking east to the cupola

68

69

67

70

Fig. 67 Original teahouse

Fig. 68 Current teahouse

Fig. 69 Cupola

Fig. 70 Chinese Chippendale bridge

71

72

73

74

Fig. 71 Small teahouse

Fig. 72 Bellevue House replica of the Colonial Revival pergola from the Sarah Orne Jewett House outside Portsmouth, New Hampshire

Fig. 73 Reconstructed pergola. The pergola was here before the 1938 hurricane.

Fig. 74 The shower arch at Bellevue House is a replica of one of the arches made by Samuel McIntire for Washington Square, Salem, Massachusetts

75

Fig. 75 Teahouse allée

Fig. 76 Volutes in the American Renaissance Water Garden

Fig. 77 Table in the American Renaissance Water Garden. Note how the drainage design is echoed in the adjacent rill.

We sought to respect the pattern language even in such modest elements as the Federal-style urn shapes of the railing ornaments on the stairs leading out to the garden on the east-facing side of the house. These urn shapes connect to the ornaments on the fences that embrace each side of the teahouse like encircling arms defining the Yew Garden; the same pattern is repeated in the railing posts on the classical chamfered granite Palladian bridge at the northern end of the emerald pool adjacent to the bathhouses (cabanas).

Similarly, we deployed volutes throughout the garden. These are the spiral scroll brackets that Renaissance and Baroque architects used to connect building levels and to support benches and foundations.[16] The volutes on the ends of the family banquet table on the terrace set the precedent for the repetition of the motif throughout the garden—in the lattice work framing the back of the terrace, on the steps of the terrace, around the fountain at the southern end of the American Renaissance Water Garden, and on the McIntire shower arch by the south pool (figs. 75–84).

Volutes also appear in the topiary lining the steps ascending from the house and behind the semi-circular bench adjacent to the children's fountain. The shapes are repeated again in the central gazebo of the allée, at the ends of the lattice on the emerald pool terrace, in the cascade leading to the nymphaeum/library at the northeast end of the property, and in the exedra behind the Neptune sculpture in the pollarded French Garden at the southwestern end of the property.

Adding to the architectural repetition, the recurring use of certain materials, particularly a rich gray-green granite, which sculptor Mark Mennin quarried on the Canadian border near Lake Placid, is another subtle element that binds parts of the garden together.[17] This material is used in the ceremonial Villa Lante-style table on the terrace, the volutes framing the steps from the terrace, and the lintels lining the rill leading down to the children's fountain. This fountain itself is carved out of a solid block of this granite, as is the semicircular bench behind it. The axis beneath the

76

77

Fig. 78 Drain in the Kitchen Garden

Fig. 79 Drain in the greenhouse

Fig. 80 Children's fountain and benches in the American Renaissance Water Garden

78

79

80

81

82

83

Fig. 81 Volutes by an anonymous artist in the *Years of Living Dangerously* cascade

Fig. 82 Mosaic in the Kitchen Garden

Fig. 83 Rill in the American Renaissance Water Garden

Fig. 84 Arts and Crafts pool

84

85

shower arch continues with the carved granite head of my son, Reynolds, in the Arts and Crafts Garden, spouting water into a basin (fig. 85). The lotus leaves in the Lutyens-style fountain at the south end of the pool are also sculpted in gray-green granite.

Trees also contribute to the pattern language. Pollarded big-leaf linden trees border the orchard to the north of the house and feature in the French pollarded garden, originally designed by Achille Duchêne, just south of it.[18] Hornbeams are also repeated: a corridor of pleached hornbeams forms the allée that now defines the south vista from the teahouse, and rows of the trees embrace the sides of the teahouse. The repetitive rigor of these trees, though expensive to maintain, adds visual pleasure to the overall garden composition

Fig. 85 Reynolds's fountain head, Arts and Crafts Garden

Fig. 86 Central allée leading north from the new teahouse through to the old teahouse and into the Yew garden

86

87

88

and helps to define the key vistas. The restored pergola leading to the McIntire teahouse repeats the vertical lines of the trees, but with columns rather than tree trunks. The columns are clad in vines and frame the view of the wooden terrace where the teahouse rests. This pattern language is supported by a system of axes and vistas that connect the focal points of the garden (figs. 86–88).

These different elements—the diminutive Federal-style buildings that terminate the vistas, the allées of trees, the carved stone work, and the pattern of lattice and volute details—all combine to create balance and symmetry in the garden. By drawing on a broad variety of references, the pattern language enlarges the cosmos of our personal garden.

Destinations and vistas

The different destinations in the enclosed world of the garden are linked by gravel pathways, now given a crisp edge with steel lining. These paths are not the road from Troy to Rome, as English banker and garden owner-designer Henry Hoare II would envision us taking to his temples and follies amidst the hillocks of Stourhead in Wiltshire, England.[19] A stroll through the Bellevue House gardens offers a different sort of journey.

A perambulation might best start in the middle of the garden at the two-story teahouse—the first significant landmark that a visitor from the 1920s would remember. It remains an orientation point today, situated on the axis that divides the garden into two parts from east to west with the pleached allée of hornbeams. A gravel walkway proceeds through the intersection of the axis, connecting the charming old gazebo built into the northern wall to the pierced wooden gate in the southern wall of the garden. This path also intersects another major east-west walkway that travels from the Codman-designed exedra at the west wall through Achille Duchêne's pollarded French Garden. The walkway then jumps the broad panel of lawn facing the south terrace of the house and continues on past the new privet-lined sunken Arts and Crafts pool and adjacent pergola leading to the one-story teahouse (fig. 89). It concludes its measured tread along a steel-edged pathway at the gazebo that defines the central axis of the garden, marked with a circular blue-green compass—the navel from whence we contemplate the cosmos of the garden (fig. 90). Standing on the compass at

Fig. 87 Central allée, viewed from the southwest

Fig. 88 Central allée, viewed from the south gate

89

90

Fig. 89 View from the Arts and Crafts pool, looking south

Fig. 90 Compass by Mark Mennin in the allée

the intersection of the two paths, the blue Brazilian quartz buttons that Mark Mennin carved are visible. To the east are mounded grass forms. The consequent enclosure of the distant emerald pool adds a sense of drama, as the large white cupola at the east end of the property can no longer be seen.

On either side of the gazebo are steel-framed trellises of apricot roses, which artist Jonathan Glatt carefully crafted, with tiny bronze capitals completing the column frames and handsome bronze feet at the base. There is an intimate pleasure in seeing the unique architectural details that scale this framing element. The trellises are capped with a keystone design in bronze that visually connects to the keystone of the arch of the teahouse and the keystone pattern of the central gazebo structure, fashioned in mahogany (figs. 91 and 92). The effect of these repeated forms is to create a short musical syncopation of beats that echo up and down this vista. It is perhaps the most sophisticated small-scale manipulation of repetitive elements in our garden.

Less architectural in feeling but still contributing to this rhythm is the pergola just west of the allée, our earliest garden project (2003)—a reconstruction of a design that old photographs revealed to have once stood on the same site; it was probably demolished in the 1938 hurricane. Instead of the trellised roses, this pergola is all columns with Laburnum (golden chain) and Allium reaching upwards, as if paying homage to garden designer Rosemary Verey's famous pergola garden at Barnsley House in Gloucestershire, England. Set on the original paving stones, the pergola leads to another McIntire-designed teahouse perched on an elevated wooden porch. It is a replica of the original one-story teahouse McIntire designed for the Elias Hasket Derby family at their city mansion, now situated behind the Gardner-Pingree House at the Peabody Essex Museum in Salem. It affords another particularly fine vantage point from which to view the new sunken Arts and Crafts Garden just to the north of it.

The most striking new garden structure and subsequent vista is the cupola folly, situated on an axis with the exedra. It is based on the design of the Branch or Howard Street Church in Salem, originally constructed in 1809. There remains only a single drawing of this church structure, which was destroyed by fire in the 1830s (fig. 93). Based on that drawing, architect J. P. Couture created a garden structure with an imagined interior that includes a hardwood-inlaid floor, recalling the pattern of the marble floor in the rotunda of Bellevue House (fig. 94). A copper monkey weathervane, fashioned by sculptor William Reimann, sits atop the folly to contemplate the six-foot wooden monkey artist Justin Gordon carved out of a dying beech tree just south of Bellevue House's graveled courtyard.

91

92

While completing the cupola, we noticed that the hard clay that lay beneath the grass there held water after rain, creating a temporary reflection of the cupola on the lawn. The reflection led us to the notion of constructing a water garden that would perpetuate this image, just as panels of water reflect follies set on the beveled grass and stone-edged shoreline of Studley Royal Park in North Yorkshire, England. We installed large granite steps running parallel to the cupola that could be joined by a wall, which would define a shoreline with the desired crisp edge. The brightly colored carp that we had expensively imported into the pond delighted small children by nibbling at their feet. The first school is gone now, as watchful herons apparently unaware of our design intention decided they wanted a free sushi lunch, but it has been since sometimes restocked.

Fig. 91 Arched trellises designed by Jonathan Glatt in the allée

Fig. 92 Inside the new teahouse, looking south

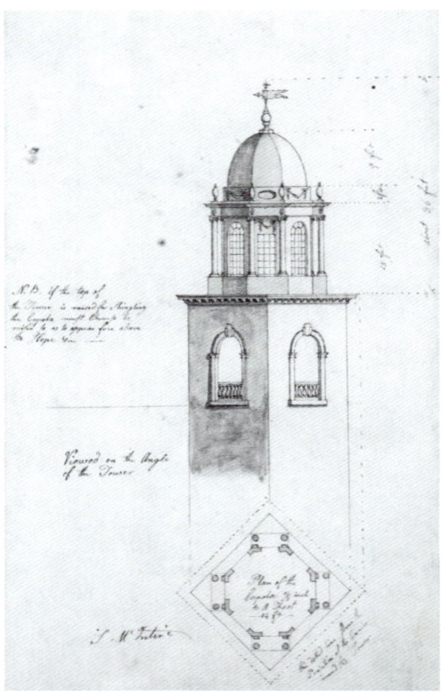

93

94

Fig. 93 Samuel McIntire, alternate elevation and plan for cupola of Branch or Howard Street Church, Salem, 1804–1805, Pen and ink and wash on paper, 38.1 × 11.98 cm (15 × $4\frac{18}{25}$ in.), Phillips Library, Samuel McIntire Papers MSS 264, Box 3, Folder 5, #60

Fig. 94 Drawing of Bellevue House cupola by J. P. Couture

Fig. 95 Chinese Dairy at Woburn Abbey and Gardens, Bedfordshire, England

The pool is now symmetrical, fed by a little cascade that is framed by a Chinese Chippendale bridge in vermillion red. The English interest in Chinese folly structures stimulated the eighteenth-century architect William Chambers to produce some designs, which led to the Chinese pavilions found at Biddulph Grange and at Woburn Abbey (fig. 95).[20] Our Chinese bridge, designed by J. P. Couture and built by carpenter Derek Kosciuszko, links the "peninsula" shaped by the edge of this formal pond and the adjacent lagoon, named the Oriental Vale (figs. 96 and 97).[21]

95

The Oriental Vale garden was inspired by a lone pine (which died) at the southeast corner of the property and three Japanese maples. Desiring to capture a reflection of the trees' leaves in the fall, we created a sinuous lagoon bordered with grasses and lily pads, where water flows over a tiny cascade into the more formal English-style pond. To the southeast we constructed a hillock with a waterfall that courses down next to the water-loving willow that replaced the pine tree—removed when it became waterlogged—thus creating the "vale." Two connecting streamlets mark this waterfall, which rises roughly three and a half feet above the garden floor, forming a landscape backdrop on the southern edge of the property. The stream and subsequent waterfall is traversed at the top by a Japanese-style stone slab bridge. At the water level, stepping-stones cross the stream as it empties into the lagoon.

A path marked by carved granite Japanese lanterns circles the new willow tree and connects it to the peninsula, leading back to the Chinese Chippendale bridge. Evergreens

96

97

frame the scene, and a "shoal" of dark blue-grey pebbles (imported from Peru, as it is harder now to purchase them locally) form a bold, dark pattern along the western side of the lagoon. The garden is full of drama in the spring and the fall, when first the azaleas and then the grasses and maples planted near the stones turn a contrasting orange and red. This is our homage to the Katsura Imperial Villa in Kyoto, with its beds of pebbles framing a lagoon, one of the most distinctive uses of these pebbles in Japanese garden design (fig. 98).[22]

The Oriental Vale is intended to resemble something that an English squire might have created in his park. Aside from the allusion to the Katsura Villa's pebbled shore, we shaped the vale with an overall effect of Orientalism, rather than

98

Figs. 96 and 97 Oriental Vale, Bellevue House

Fig. 98 Katsura Imperial Villa, Kyoto, Japan

99

100

101

102

specifically replicating any particular Japanese or Chinese garden. In the fall, the three maples define its southern edge with the copper sheen of their leaves, and from the bridge, one can look down upon multicolored carp (sometimes) swimming in the pond. The Oriental Vale is located adjacent to the southern edge of the property and serves as a gift to the street, pleasing neighbors and passers-by through (ever-diminishing) holes in the arborvitae.

To the north of the cupola and on an axis with the pond lies the emerald pool, hidden behind berms so it remains a surprise awaiting visitors coming from the house (figs. 99–102). It is framed in the rear by a stone terrace flanked by bathing pavilions, which vaguely recall a garden house that Samuel McIntire designed for The Vale, the Lyman family estate in Waltham, Massachusetts. They also resemble Bahamian cottages, as they are tiny in size with outdoor showers behind them. Between the little houses stands the Family Chimney Tree.

A stroll through the Bellevue House gardens, then, if not Hoare's journey from Troy to Rome, is perhaps akin to an exploration of the grounds of Quinta da Regaleira, on the outskirts of Sintra, Portugal. The gardens, designed in 1898 by owner António Augusto Carvalho Monteiro and the Italian architect and set designer Luigi Manini, who began his career at La Scala in Milan before moving to Lisbon, feature a startling range of structures: these follies, towers, belvederes, staircases, gateways, fountains, and naturalistic pools and cascades are the focal points of a series of carefully composed sets that lie concealed among the trees, each reflecting Monteiro's philosophical beliefs. Like the Bellevue House garden, Quinta da Regaleira has a cosmology and employed numerous artists and artisans throughout its construction. This is the most provocative example of a narrative garden that we have discovered that might connect to our own efforts.

> The garden, as an image of the Cosmos, is revealed through a succession of magic and mysterious places. The quest for paradise is found in coexistence with a mundus inferus—such as Dante's Inferno—through which the candidate for initiation is led by an Ariadne's thread. Various scenes from the initiate's journey appear along the way of the vera peregrinatio mundi through a symbolic garden where we can feel the Harmony of the Spheres and examine the perspective of an ascetic conscience, by analogy to the metaphysical quest for the Being that is found in the great Epics. In these realms abound references to the worlds of mythology and to the mission of the Templars, to great mystics and miraculous magicians, and to the enigmas of the alchemical Ars Magna.[23]

Apparently, the King of Portugal, who resided across the street, never saw the garden of Quinta da Regaleira. We are more optimistic about visiting prelates.[24]

Fig. 99 Alignment in the American Renaissance Water Garden

Fig. 100 Children's fountain in the American Renaissance Water Garden

Fig. 101 Emerald pool

Fig. 102 Arts and Crafts pool

Notes

1 Pauline Metcalf (editor of *Ogden Codman and the Decoration of Houses*), interview by author, Newport Garden Show, Newport, Rhode Island, 2015.

2 Until recently, Derby's connection with the house was unknown, except to the head gardener, Horus, who discovered it on Ancestry.com!

3 Dean T. Lahikainen, *Samuel McIntire: Carving an American Style* (Salem: Peabody Essex Museum, 2007), 39.

4 International Council on Monuments and Sites (ICOMOS), *International Charter for the Conservation and Restoration of Monuments and Sites* (The Venice Charter), 2nd International Congress of Architects and Technicians of Historic Sites, Venice, 1964, https://www.international.icomos.org/charters/venice_e.pdf.

5 ICOMOS, Venice Charter, article 3.

6 Ibid., article 9.

7 Professor Steven Semes provides the best theoretical analysis of the disconnect between modernist architecture and preservation practice in *The Future of the Past: A Conservation Ethic for Architecture, Urbanism, and Historic Preservation*, The Classical America Series in Art and Architecture (New York and London: Norton & Company, 2009).

8 Ronald Lee Fleming, *Saving Face: How Corporate Franchise Design Can Respect Community Identity*, 2nd ed. (Chicago: American Planning Association, 2004); 1st ed. 2002.

9 See National Park Service, U.S. Department of the Interior, *The Secretary of the Interior's Standards*, https://www.nps.gov/tps/standards.htm.

10 Brent C. Brolin, *Architecture in Context: Fitting New Buildings with Old* (New York: Van Nostrand Reinhold, 1980).

11 Codman gave his historical collection of architectural and garden design materials to the Avery Library at Columbia University.

12 The trellis now stands just behind the high brick wall that defines the northern boundary of the property. These walls separate Bellevue House from the street; across that street is the long, low-slung 1950s concrete-block shopping center that begins the commercial section of the Bellevue Avenue district, formerly the grass courts of the Newport Casino, and before that the Ocean House Hotel.

13 "Blowin' in the Wind," track 1 on *The Freewheelin' Bob Dylan*, Columbia Recording Studios, 1963, vinyl.

14 Fiske Kimball, *Mr. Samuel McIntire, Carver: The Architect of Salem* (Portland, ME: The Southworth-Anthoensen Press, published for the Essex Institute of Salem, 1940). Mrs. Endicott and her son, William Crowninshield Endicott Jr., salvaged the original teahouse and reconstructed it on their estate Glen Magna Farms, in Danvers, Massachusetts, in 1901. The teahouse still stands in their garden today, not far from the site of the current Northshore Mall where it originally stood, and now visible from Interstate 95, which traverses Danvers, when the trees are bare of leaves in winter.

15 Entirely new spaces include the allée garden, the emerald pool, the Arts and Crafts Garden, the American Renaissance Water Garden, and the Oriental Vale.

16 John H. Parker, *The Concise Dictionary of Architectural Terms* (New York: Dover Publications, 2011).
The word *volute* derives from the Latin *voluta*, meaning "scroll." It has been suggested that the ornament was inspired by the curve of a ram's horns, or perhaps was derived from the natural spiral of the ovule of a common species of clover native to Greece. Alternatively, it may simply be of geometric origins.

17 Mennin usually does more abstract and personal work, but recently he has designed sculptures in the classical tradition.

18 Achille Duchêne (1866–1947) was a French garden designer who worked in the grand manner established by André Le Nôtre. He was responsible over a period of years for some six thousand gardens in France and worldwide. Notable examples of his work in France include the restoration of the gardens at Vaux-le-Vicomte, Versailles, and the Château de Courances, and in England the creation of the Water Parterres at Blenheim Palace (financed by a Vanderbilt dowry to the Duke of Marlborough).

19 Henry Hoare II (1705–1785), known as Henry the Magnificent, was an English banker and garden owner. Henry dominated the Hoare family through his wealth and personal charisma. His nickname derived in part from his influence as a great patron of the arts, but more particularly from the gardens he laid out at his Wiltshire estate Stourhead, which became a masterpiece of English garden design. The garden, in the school of Poussin, was said to be "more beautiful than any landscape put on canvas." The gardens were admired as a showpiece. Capability Brown (1716–1783), the renowned English landscape gardener, was well known to Hoare. See Lydia Greeves, *History and Landscape: The Guide to National Trust Properties in England, Wales and Northern Ireland* (London: National Trust Enterprises Ltd., 2004),

366–69 (Stourhead), 374–75 (Stowe), 425–26 (West Wycombe Park).

20 Sara Burdett, *Biddulph Grange Garden*, National Trust Guidebooks (Warrington, England: National Trust, 2010); Keir Davidson, *Woburn Abbey: The Park and Gardens* (London: Pimpernel Press, 2016).

21 William Chambers, *Designs of Chinese Buildings, Furniture, Dresses, Machines, and Utensils: To Which Is Annexed a Description of Their Temples, Houses, Gardens, &c.* (London, 1757); William Chambers, *A Dissertation on Oriental Gardening* (London, 1772).

22 Tadashi Ishikawa, *Imperial Villas of Kyoto: The Katsura and Shugaku-in*, 2nd ed. (Tokyo: Kodansha International Press, 1970).

23 *Quinta da Regaleira*, brochure (Sintra, Portugal: Cultursintra Foundation, 2013).

24 Helena Attlee, *The Gardens of Portugal* (London: Frances Lincoln, 2008).

TRAILS
COLLEGIATE
Redlands Smiley Library
Virginia Dare Winery
66
First McDonald's
Redlands Railroad Station
Mt. Wilson Observatory
Pitzer House
Frary Dining Hall
College Gates
POMONA COLLEGE
California Institute of Technology Athenaeum
Little Bridges
Greek Theater
Gamble House
Claremont Inn
Riverside County Courthouse
Los Angeles County Fairground
Santa Fe Railroad
Kellogg Ranch
Pilgrim Congregational Church
PACIFIC ELECTRIC
CLAREMONT
AUCTION SALE

Chapter Two

THE PLACE OF MEMORY

Resonating in the soul

Chapter 1 identified the framework of design elements that configure the garden space and the pattern language of design features that provide the continuity of texture. Landmarks define the spatial relationships in the garden terrain, the topography that explorers can use to orient themselves. In anatomical terms, this physical landscape is the bones of the garden. By the same analogy, in this chapter we look at the connective tissue that binds place and memory together to recapture accumulated experience.

To some extent, the recollection of activities in garden spaces builds up a resonance in our imagination. The shapes and forms of design features are physical cues that stimulate remembrance of past events; they are manifestations of the external events that shape personal history. How, then, to embed this sense of recollection and subsequent resurrection of memory in a garden? Just as a madeleine prompted Proust to recall his childhood visits to have tea with his aunt in Combray,[1] so the design elements embedded in a garden can restore and recapture both the external reality of the garden and the mental sensation of experience in all its fullness.

Fig. 103 Cultural landscape of Pomona by Gregg LeFevre, bronze, Pomona College, California

Cultural landscapes

The earliest evocation of place in the Bellevue House garden is a "cultural landscape" of two valleys in Southern California, cast in bronze relief by the New York artist Gregg LeFevre. These panels were originally designed for architect Robert A. M. Stern's open atrium at the Smith Campus Center at Pomona College, but a duplicate found its way into the Yew Garden behind Bellevue House. Eight family members attended Pomona College, and the bronze scene frames the campus they remember in a broad hundred-mile band of valleys. The landscape stretches from the Riverside Inn and the Redlands Library in the east all the way to the Arroyo Seco in Pasadena to the west.

The major buildings of the college are clustered in the middle of the sculpture, framed by a circle of mementos of Pomona College history mediating the surrounding cultural landscape. Indian baskets, the college humor magazine, and a Santa Fe railroad real estate promotion map connect with the cast designs of orange crate labels and smudge pots, reminders of the orange belt that once embraced the garden cities of the Pomona and San Gabriel Valleys and infused them with perfumed springtime blossoms. There are images of the token for passage on the Red Cars of the Pacific Electric, at one time the most comprehensive rapid transit system in the country, with a network covering two thousand miles around the Los Angeles basin. The college mascot, the sage hen (a type of grouse), is also featured. The

103

104

Fig. 104 A replica of Gregg LeFevre's bronze cultural landscape now sits in the Yew Garden, Bellevue House

first half of the dedicatory inscription beneath the landscape recalls a time, a place, and a civic view that has been eroded, just as the landscape beyond the college campus has been eroded, losing its own connectedness in a morass of suburbia. Perhaps the area lost its bearings as the City Beautiful movement lost its force in American culture.[2] Was it the last stand or high-water mark of a certain high WASP culture? The legend on the bronze relief begins as follows:

> The terrain stretches from Arts and Crafts bungalows on the Arroyo Seco's rim to the invented mythos of revival styles at Riverside's Mission Inn. The bronze work also evokes the elusive memories of a lost Eden. Sprawling malls and ranchettes have vanquished shimmering eucalyptus on country roads and scented orange groves once plaiting the garden cities of the Pomona and San Gabriel Valleys. Yet an early and generous civic imagination still enriches the axis of this landscape. The Romanesque Smiley Library in Redlands, the greensward down Euclid Avenue in Upland, and the Baroque dome of Pasadena's City Hall recall the idealistic vision of the City.
> *Ronald Lee Fleming* [3]

This 12-by-12-foot bronze square now intersects the path connecting the original gazebo in the north wall of the Yew Garden to the two-story McIntire teahouse, the oldest landmark structure in the garden (fig. 104). The bronze landscape rests between the two circles of the giant Irish yews and mediates between the gazebo and teahouse. It forms a contemplative space, framed in boxwood on the axis between the two structures, and enriches this short passage. Unlike the old square-dancing tune, the circle never directly meets the bronze square in this design. It transcends the walled garden space and contemplates the memory of a world encompassed by two valleys where our family experienced a sense of place for more than a century.[4]

Imagining a journey through time

One of the key, though perhaps subconscious, influences that shaped the narrative of our garden was Holling Clancy Holling's books *Seabird* and *Tree in the Trail*, which we had read as young children and which our whole family adored.[5] Both books engage young readers with beautifully illustrated stories of the history associated with artifacts from the past. *Seabird* follows a sailor's carved ivory bird through time, from clipper ships and whaling vessels to modern ocean liners; the context changes, but the carved ivory object remains the same. *Tree in the Trail* explores how experience can transform an object. It tells the story of a young sapling on what becomes the Santa Fe Trail. The young tree offers shade to travelers passing by; a boy protects the base of the tree with sharp stones so it can survive and grow. Native Americans, Spanish conquistadors, mountain men, frontiersmen, and cowboys all leave their marks on the tree's trunk and branches, from obsidian arrow heads, to iron lance points, to rifle bullets that pierce its bark, until eventually the old tree dies. A passing drover salvages a piece of the wood and carves it into a yoke for his oxen, skillfully commemorating the tree's life with embedded brass tacks that display the mementos of the passing years.

Tree in the Trail was an early placemaking influence as it built for me an association between place and time, between history and context. It encouraged me to eventually theorize about a book of placemaking design as a function of my own garden making.[6] It was probably also an influence on the habit I developed in childhood of collecting artifacts, beginning with the souvenirs of my grandparents' 1922 cruise around the world, which I discovered at the age of four.[7] Later I acquired antique weapons and old books on historic towns, built a rock collection, and searched for postage stamps, all of which became props for my imagination.

When I was ten, my parents had a friend from church who was a curator at the Los Angeles County Museum, which held General Patton's collection of weapons, along with other firearms and edged weapons not on public display. I would meet him at the museum on Saturdays and he would lock me in a storage room with bushel barrels of muskets and samurai swords, German spiked helmets, even a brass cannon or two. It was heaven. I even made a museum of my own in my bedroom, which was an obligatory visiting point for my parents' bridge partners and square-dancing crowd.

My parents liked old things, and our house glowed with the patina of antiques, but they were not obsessed as I was. My serious addiction of topophilia, or love of place,[8] resulted from family visits to ghost towns all over the West, from the abandoned gold-rush camps of the Mother Lode in California to the empty silver-mining towns in Nevada and Colorado. From Bodie to Rhyolite, from Calico to Virginia City, my parents made these dusty treks with a cajoling and tenacious son in the back seat of the family car. I can still remember my father's fidgeting reluctance as I demanded that he stop the car to read yet another roadside historical marker. (My sister rocked back and forth in the back seat and wanted no part of this!)

I became fascinated by landmarks, the German word for which is *Denkmal*, meaning "time stop" or, more poetically, "reflect for a moment."[9] This fascination took its most extreme form in the constructing of my own ghost town in the empty lot adjacent to my parents' house in a hillside suburb of Los Angeles County. This lot helped to fix those landmarks in my imagination, aided by a little library of books that I collected on the history and condition of old mining towns. That tiny library became the foundation for the seven thousand books now in the library at Bellevue House, which has its own building.[10]

After studying old photographs and visiting the sites, I gradually constructed a "main street" in my ghost town. First came a sheriff's office, a trim and plumb structure that my father built—in the Western films I watched, the lawmen always won against the Indians. Then followed more ramshackle structures, salvaged in part from packing crates, where neighborhood boys playing lawmen assembled under my Tom Sawyer-like direction. (My sister was sometimes on the losing side as an Indian. This may have left lasting resentment.)

The ghost town grew to include a general store, Hangman's Hotel (complete with a swinging carved wooden sign), and mining shacks. Two boot hills were planted with wooden grave markers, their epithets vividly describing an imagined cast of nefarious characters who usually met their end in a blaze of cap-gun fire on our street. A stream bed and wagon track encircled the town. I planted iris along the stream bed (my first gardening attempt). I let the water from the faucet on the upper flank of the hillside site run for days at a time, and it coursed its way down the hard pan that passed for soil past little dams to a vegetable patch. Though water was a precious commodity in thirsty Southern California, my parents were not known to raise their voices in protest. Looking back, it seems an enduring testimonial to their support (and their forbearance).

Much later, while studying at the Harvard Graduate School of Design, I discovered that my ghost town was actually an "adventure playground." The pioneering authority on such playgrounds, the formidable Lady Allen, was influential in defining this in the literature of the day.[11] Of course, my parents took no notice of the literature; they only cared that their son and heir and his more reluctant little sister were out of harm's way, fully occupied at the ghost town, where neighborhood children came to play with us, and not smoking cigarettes in a back alley.[12]

When I was in the eighth grade, my grandmother moved from Hermosa Beach, wanting to be closer to my mother, her only child, and my parents decided to build a house for her on the lot next door. One morning I awoke early in horror to see a bulldozer advancing on my ghost town. Within a short hour or two it had been reduced to a pile of kindling. I was a minor victim of urban renewal, a strategy that swept through American cities and towns and mowed down many livable and lively neighborhoods, sometimes replacing them with single-use structures and sterile blocks for people other than the original inhabitants.[13] This phase in urban development reduced the language of many places, kicking the teeth out of many neighborhoods and leaving gaps in the smiles of so many handsome districts.

Reclaiming lost landscapes

"American cities change faster than their inhabitants do," Jean-Paul Sartre commented when he visited America in the 1950s.[14] The notion of reconstructing the lost landscapes of cities in a country where so many communities destroyed their architecture before they were able to appreciate what they were losing became a theme in our own garden, which commemorates Southern California landmark structures. The bronze map of the "cultural landscape" recalls to us the locations cherished by our grandparents and great-grandparents but now lost in the surrounding cityscape. Some of this landscape does still exist, but it is often obscured behind the sprawl of anonymous suburbia. After returning from Vietnam, I decided to take an old Wellesley girlfriend to visit my grandmother's house in San Marino. We never found it; I completely lost my bearings in the bewildering face of a new development.

The early Spanish land-grant ranchos, low-slung adobe constructions, are often still there; the old missions like San Gabriel are easier to find. But the beautiful gardens in Redlands, built by the great-grandparents of a college classmate, were destroyed because the city did not accept them when the Smiley family offered them as a gift (fig. 105).[15] The family still owns the Mohonk Mountain House in New Paltz, New York,[16] where the Quaker founder of this celebrated hostelry used to organize international peace conferences; they also gave the funds for Smiley Hall at Pomona College and Smiley Library in Redlands, now one vignette in our bronze cultural landscape.

The bronze map at Pomona College allows undergraduates and their parents to trace the outlines of the landmarks under their feet. Successive presidents of the college have told me that parents sometimes stand on the map for twenty minutes or more, exploring the topography. As children driving with our parents to football games at Pomona College, which my parents and my mother's cousins and cousins-in-law had attended, we watched much of this landscape disappear. In those days there were upright Victorian houses with their skirts of gingerbread siding still sitting in orange groves along the road to the college. Then came the freeways; bulldozers mowed down the dying groves and the land lost its springtime fragrance of orange blossoms. Those houses that survived appeared as ravished matriarchs, shorn of their Victorian painted finery, their windows hollow-eyed, ragged curtains flapping in the breeze. Little ranchettes—houses made of "ticky tacky," as the song goes—appeared as we were going to college.[17] The street car lines disappeared too. Once they brought my grandparents from Hermosa Beach to the headquarters of the family jewelry business, E. W. Reynolds Company, in downtown Los Angeles in twenty-five minutes. Now ever-expanding freeways, clogged to the gunnels in rush-hour traffic, imprisoned commuters in heavy congestion for much of the day. We had to pick our hour carefully when returning to the college.[18]

105

Fig. 105 Smiley residence, Canyon Crest Park, Redlands, California, c. 1900

The removal of landmarks in Southern California shaped my experience, and so the idea of charting memories in the garden seemed a way to reclaim that lost landscape. Some cities publish "lost landmarks" books, but many lose their historic architecture before the community's consciousness is strong enough to even understand what they have lost or to document it with a book. Oklahoma City, for example, lost many Art Deco landmarks before there was a critical understanding of how important they were. At least Tulsa published a book![19]

The fragile terrain of family memory

As the Bellevue House gardens started to assert their own identity in their interlocking green rooms and spaces bisected by leafy passageways, and their powerfully composed vistas of classical buildings, the idea of a family narrative—the legacy of our particular family's dance through the passage of time—became harder to register, much less preserve. Now, after twenty years of planning and building, there are powerful images in the garden that can manifest their claim on our memory. People visiting the gardens today are able to form their own associations with these images based on their own experience, which is often amplified by animation events (see chapter 3).

Evoking family memory is more challenging now because family memory is more fragile. The context of our family's experience is cast in a bronze map, and illustrated in bronze vignettes wrapped around the Family Chimney Tree (which rises above a pizza oven at the children's pavilion and cabanas around which the family congregates), but do these powerful images actually register on a new generation that hasn't heard the stories or exercised their own imagination in long-cherished rituals at Bellevue House?

The vignettes on the Family Chimney Tree depict ancestors who connect this family to the Society of Colonial Wars and, through Captain Giles Mead, to the Society of the Cincinnati (fig. 106). They chronicle battlefield death at Chancellorsville (a Reynolds ancestor) and a battlefield commission at Shiloh (a Fleming ancestor). They depict racing buckboards at the Oklahoma Land Rush and a marshal's badge in the San Francisco earthquake and fire. They show my narrow miss by a sniper's bullet during the Tet Offensive in Saigon.[20]

Fig 106 Detail of the Fleming Family Chimney Tree, Bellevue House

Our family tree may record names, events, and dates, but they are far less permanently etched in the memory in the next generation. They are at best tenuous and oftentimes illusive; they fall victim to the territorial dislocation of children, the emotional cauterization of divorce, and even different teaching methods in high schools, which sometimes dismiss chronological history for the study of particular issues. Who now remembers our ancestor, Captain Giles Mead, who escorted the French troops from Newport to Yorktown? We didn't! Our family history disintegrates even as it is contoured in a bronze map and sculpted in bronze vignettes. This is perhaps the greatest irony confronting, even haunting, a narrative garden builder.

Thanks to the power of the internet and genealogical resources like Ancestry.com, it is easier to connect the chronological family dots. We have been able to create a framing trunk on either side of the vignetted branches of our family tree: one trunk for the Reynolds family, who arrived in Watertown, Massachusetts, in 1634; the other for the Flemings, the younger sons of a Scottish laird, who arrived at Harrison's Landing, Virginia, near Williamsburg, in 1619.[21]

Each family had their symbols of endeavor and each had an institutional memory that was broken over time. In the last several generations, however, there have been only nuclear families, and no grandfathers around to tell stories of early exploits as the Flemings moved by increments across the land: Virginia, North Carolina, Kentucky, Ohio, Indiana, Iowa, Kansas, Nebraska, Oregon, California, and Massachusetts. The Reynoldses—with more memory intact thanks to the record of an invalid's passage by train from Stamford, Connecticut, directly to Southern California—kept more of their Yankee customs and a stronger sense of their lineage. With the prosperity of great-grandfather E. W. Reynolds, a successful merchant who founded what was at one time the largest wholesale jewelry business west of Chicago, they retained a family pride of place with houses in the Wilshire district and Beverly Hills, clubs and camps at Hermosa Beach, and later a house at Smoke Tree Ranch in Palm Springs (the Hobe Sound of the West). Eventually, however, the other side of the Reynolds family moved the business headquarters out of the jewelry district and went bankrupt because of it. As some families become wealthy, they can lose their understanding of the family context and thus their survival genes.

The conjoined Reynolds/Fleming family, now also merged with a Swiss family, had deep linkages and connections that were shattered by the divorce of my wife and I when our children were very young, making it harder to conjure up the distant family chronologies. Though there were family images dating back to the first Californian immigrant in the 1850s, there was no telling of tales on a grandfather's knee, no ancient boxes of letters evoking the family's history through time from the seventeenth century. Indeed, my children and I have clearer memories of the Thomas Boylston Adams family in Lincoln, Massachusetts, with whom we spent Thanksgivings and Christmases after the divorce.[22]

Perhaps this is why the garden at Bellevue House becomes more urgent, more significant, and ultimately more poignant to me. Through it, I want to make vivid again this personal memory of reconstituted linkages, even to visitors who remain largely unconscious of our family story. Most people come to the garden for music making, dancing, daffodil gazing (we have about fifteen thousand planted), reverie, and general delight; they have no particular interest in either Flemings or Reynoldses, nor in our ancestral lines. As the grounds are given over to civic merriment, will there be enough evidence to engage visitors in the story of an early family founded, then lost, then found again (maybe), while the band plays on for casual strollers? Will they return to cherish the garden and its story anew? Will they replenish their own rituals from a well of diminished memory? Will they build enough new memories here and now to cherish our garden place as common ground?

107

108

Cascading personal memories

A series of little sculptures and a cascade near the Family Chimney Tree extends the sense of reflection that the depiction of ancestral lines is designed to evoke. It asks that we take full measure of the dangerous passages in our lives, to be present, to be aware, to better understand risk, and to celebrate survival. Entitled *The Years of Living Dangerously*,[23] the cascade is different from the runnel in the American Renaissance Water Garden that traces the passage forward from father to son to grandchildren; it is shorter and metaphorically more dangerous, and rushes from a quiet pool past death-defying obstacles to the eventual paradise of the nymphaeum below the rushing stream (figs. 107–11).

The cascade appears to originate in a calm emerald-green pool, from where it then flows under a chamfered granite bridge of classical proportions to close the garden's northern vista at the bridge portal. Here, the water literally thunders into a sound chamber and is channeled into a passageway that visually connects the bridge to the nymphaeum entrance. A slightly elevated pavement in the stream bed creates ripple effects. The terraces on either side of the cascade feature three dolphins, bracketed by volutes and with monkeys perched like jockeys astride them, that spurt jets of water into bronze urns set in the rippled channel.

Small sculptures, mounted on shell and topiary parterres segmented by the volute forms, represent dangerous events and potentially deadly moments in my life: almost fatal shootings or near misses in Katanga (a breakaway and short-lived state in the Belgian Congo); a sniper's near miss during the Tet Offensive in Saigon; traffic accidents or near misses in Montenegro, Morocco, and New England; and a potentially fatal stroke. Finally, at the rim of the waterfall splashing into the nymphaeum is a cast bronze kidney, representing my transplant. Each little sculpture symbolizes a crisis past which rush the turbulent waters of life, coursing against the bronze urns and then rushing forward toward "paradise"—the shell-encrusted nymphaeum under the library. The cascade memorializes blind adventure, youthful folly, stupid choices, narrow escapes, and lucky misses. Perhaps these trials have a Bunyonesque simplicity; they are certainly not as grandiose as the seven labors of Hercules.

Embossed words on the glass panels of the bridge may recall the artist Ian Hamilton Finlay's garden of Little

Fig. 107 Library, stable block, and greenhouse

Fig. 108 *The Years of Living Dangerously* cascade, facing west

109

110

Fig. 109 *The Years of Living Dangerously* cascade, facing south toward the cupola

Fig. 110 *The Years of Living Dangerously* cascade, facing south

Fig. 111 *The Years of Living Dangerously* cascade, facing south under the bridge

111

Fig. 112 Little Sparta, the garden created by the artist Ian Hamilton Finlay in the Pentland Hills, Scotland

Fig. 113 Bronze skull by Gregg Lefevre and Jennifer Andrews in the *Years of Living Dangerously* cascade, Bellevue House

Fig. 114 Upside down Volkswagen by Gregg LeFevre and Jennifer Andrews in the *Years of Living Dangerously* cascade

Fig. 115 Sculptures by Gregg LeFevre in the *Years of Living Dangerously* cascade

112

Sparta in the Scottish Lowlands (fig. 112).[24] This is a garden of charming juxtapositions: text carved in stone, sculptural fragments, moss, and rivulets of diverted stream water. Its narrative is certainly less explicit (and, I think, more enchanting) than that of the Bellevue House gardens, though ours has a certain force of exposition. While Finlay's garden includes replicas of classical sculpture and even a tiny model of a dreadnought, our imagery is more direct and personal: a Katanga flag; a hole in a windshield of a Jaguar XKE; a bullet suspended three inches next to a skull (fig. 113); another skull cracked open, symbolizing a subarachnoid hemorrhage; two VW Bugs representing roadside near-disasters (figs. 114 and 115), one involving driving up a snow bank, falling upside down, then bouncing up again, the other a skid three hundred feet above the Adriatic (no guard rail). Our sculpture is more tightly confined than that of Little Sparta, but we hope it releases powerful associations. Ultimately, these memory pieces are designed to set a mood and to encourage reflection. They may call to mind the words that Paul Gauguin left behind in what he considered to be his masterpiece, now at the Boston Museum of Fine Arts: *D'où Venons Nous / Que Sommes Nous / Où Allons Nous* (Where do we come from? What are we? Where are we going?)

While these works of remembered peril are unique to this garden, the concept of narrating family experience can be adapted to many gardens wherever people remain concerned with questions of identity and life passage. The gathering up of memories, more permanent than the raking of leaves or the cutting of bouquets, should add layers to the narrative of the garden. These memories, now cast in sculptural form, add a third dimension to our memoir of place: reimagination. We hope they can add a depth to the powers of perception, a sort of "what if" moment. Perhaps, upon reflection, they add a certain poignancy, even redemptive grace, as we contemplate our own mortality; so we are framed—and being framed—by little sculptures and rushing waters that mark the speed of life going by.

113

114

115

Evoking historic gardens

Moving beyond personal and family memory, we also sought to achieve a certain resonance between gardens that were internationally recognized and our own modest efforts on the three-and-a-half-acre site at Bellevue House. This was not slavish imitation, but an impulse to discern and distill key design ideas and to connect them to our garden plan in way that might prompt a sense of recognition in visitors who have viewed or studied other gardens, in garden books or on garden pilgrimages.

Of course, these efforts are not always successful. Such a strategy can create amusing juxtapositions of scale as we attempt to distill elements of grand historical gardens into spaces that are often minute compared to their original settings. For example, the garden designers we hired to create the English pool adjacent to the Oriental Vale did not understand the formalism that we hoped to express in referencing the lazy river of slowly moving water skirting the classical temples and *tempiettos* at Studley Royal Park in North Yorkshire, a World Heritage site (fig. 116).We determined that the essence of this design was a crisp upright edge to the limpid waters, so the composition consisted of a plane of water, the edge of a low riverside stone wall, and a panel of grass on a bank occasionally interspersed with upright classical forms. It was a very satisfying composition. Unfortunately, our garden designers did not have that same image in their mind's eye. As they were not aware of the historical context, they made a rock-strewn edge, in a rustic style that missed the entire point of the sculpted landscape and architectural forms—that sharp demarcation at water's edge makes the composition. But we didn't show them a photograph; they just didn't get it.

We did manage to install some stone steps that run parallel to the cupola on the north shore of the pool (fig. 117); we want to further extend this shoreline with a wall that circumvents the pool edge to attain the crispness of form that we so admired at Studley Royal. The brightly

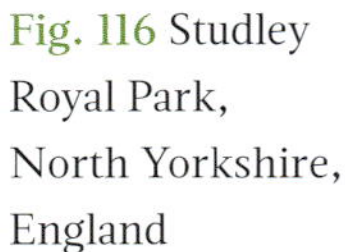

Fig. 116 Studley Royal Park, North Yorkshire, England

Fig. 117 Cupola, Bellevue House

Fig. 118 Katsura Imperial Villa, Kyoto, Japan

Figs. 119 and 120 Oriental Vale, Bellevue House

116

117

colored carp, which we had expensively imported into the pond, were a delight to children sitting on the steps; they enjoyed the nibbling of the toothless carp on their bare feet and the flicker and flash of orange through the water. The carp are gone now; the watchful heron disregarded our design intentions in favor of a free sushi lunch.[25]

We were more successful in our design citations in the adjacent Oriental Vale, where a tiny shoal of blue pebbles echoes the larger peninsula of stones at the lakeside Katsura Imperial Villa in Kyoto (fig. 118). This seventeenth-century villa is ironically one of the great icons of modernism, with its spare backdrop of elegant screens, and its walls of abstract white panels. Modern designers salute its purity of intention and timeless quality.

We brought more Asian influences into the garden with an intricate Chinese-style bridge, based on a design by the eighteenth-century English architect William Chambers. Chambers designed chinoiserie for the English gentry. His forms simplify a more elaborate Chinese Chippendale style, so his designs are compatible with the classicism of our follies. The bridge creates a foreground in the walkway progression to our rotunda. J. P. Couture incorporated bells into the bridge design to suggest Chambers's four-poster Chippendale beds, which he festooned with little bells (figs. 119 and 120).

We have also cultivated similarities between Edwin Lutyens's Hestercombe (fig. 121) and the Bellevue House Arts and Crafts Garden with its basin beneath an arched and keystoned cove (fig. 122).

The sapphire pool receives the waters of our Arts and Crafts basin. This pool, deliberately shallow so our grandchildren can more easily splash in its waters, carries the eye southward from the basin. The Gertrude Jekyll-style urns along the edges of the pool,[26] the broken stone pavements in the culminating sequence around the fountain at its terminus, and the millstones and plant pockets inserted in the stone walls, all invoke the Arts and Crafts style. The southern end of this garden, featuring a Lutyens-inspired fountain in gray-green granite, is intended to evoke, in miniature, the enormous scale of Lutyens's most ambitious garden behind the vast Rashtrapati Bhavan (formerly the Viceregal Lodge that he designed in New Delhi (figs. 123 and 124). Some of the gardens that influenced the Bellevue House narrative—like Lutyens's—are so large in scale that it may be a stretch for visitors to understand the comparisons, but we can create echoes through particular forms (figs. 125–28).

The problem of finding a scale that is easily relatable to our own gardens is also demonstrated by La Foce in Italy (fig. 129). La Foce is a vast estate, owned by English-born writer Iris Origo, with gardens designed by Cecil Pinsent, an Italophile like his forbearers Peto and Lutyens. La Foce has a viewshed anchored by carefully sited cypresses (the frequent subject of Italian travel posters) on the hillside road across the valley. Though La Foce is simply too big for us to reference directly at Bellevue House, we saw that we could connect to it through the pergola of another garden designer we admired, Rosemary Verey (figs. 130–32). Her garden at Barnsley House in England is modest but recognizable across the garden world, and her pergola, with its drapery of wisteria and laburnum, was perhaps influenced by Pinsent and Origo a generation earlier. Verey's autographed books were gifts to my daughters, who could view Verey's rustically inspirational small enclosure and then better understand the meandering and extended pergola at La Foce. Other subtle points of contact with larger gardens include water features and design details (figs. 133–36).

118

119

120

121

122

Fig. 121 Hestercombe House and Gardens, Taunton, Somerset, England

Fig. 122 American Renaissance Water Garden, Bellevue House

Fig. 123 Fountains designed by Edwin Lutyens in the Mughal Gardens, Rashtrapati Bhavan, New Delhi, India

Fig. 124 Lutyens-style fountain in the Arts and Crafts Garden, Bellevue House

123

124

125

126

127

Fig. 125 Water table, Villa Lante, Bagnaia, Italy

Fig. 126 Villa Lante-style water table, Bellevue House

Fig. 127 Getty Villa, Pacific Palisades, California

Fig. 128 Children's fountain and benches in the American Renaissance Water Garden, Bellevue House

128

129

130

Fig. 129 La Foce, designed by Iris Origo and Cecil Pinsent, Chianciano Terme, Italy

Fig. 130 Pergola, Bellevue House

Fig. 131 Barnsley House Gardens, designed by Rosemary Verey, Cirencester, Gloucestershire, England

Fig. 132 Close-up view of the pergola, Bellevue House

131

132

133

134

135

Fig. 133 Water feature at Getty Villa, Pacific Palisades, California

Fig. 134 *The Years of Living Dangerously* cascade, Bellevue House

Fig. 135 Volute forms at Hever Castle, Kent, England

Fig. 136 Volutes by an anonymous artist in the *Years of Living Dangerously* cascade, Bellevue House

136

Laying claim to place

Our garden is fragile; it is only tenderly rooted in Newport—as is our family. On the Reynolds side, we left dour New England for sunny Southern California for more than a century. A cenotaph now marks the site of our arrival with Sir Richard Saltonstall in what was then Cambridge, Massachusetts, now just east of Watertown Square. We had followed the Reverend Hooker to Wethersfield, Connecticut, in some doctrinal dispute (Puritans were always falling out in these contentious arguments). We briefly perched there, near Hartford, before floating down the Connecticut River to claim our proprietary ground in Greenwich. We intermarried there with the Mead family of nearby Mead Point, who were also among the proprietary families, numbering fewer than a score, on Greenwich Common.

Captain Giles Mead, as a commissioned officer in George Washington's army, became our propositus for membership in the Society of the Cincinnati. As a twenty-nine-year-old, he escorted a French general of the same vintage from Newport's rocky shore to the siege site at Yorktown. This general was Count Thibaut de Ménonville, who was later wounded on the redoubt there. Coincidentally, at least two hundred thirty years later, a direct descendant of General de Ménonville traveled to Newport to celebrate an engagement to my daughter Siena. He stood at the rotunda overlooking what would become the Oriental Vale and said, "We have been here before." When I expressed incredulity, he said, "Yes, we came with Lafayette on *L'Hermione*." A friend, also in the Society of the Cincinnati, heard this and confirmed, three weeks later, that our ancestor Giles Mead had escorted those French troops. Later at the destination wedding in Sicily, which I could not attend because of a pending kidney transplant, my son, Reynolds, gave a toast that evoked this union of families after so many years.

This story speaks to my feeling that families should claim particular places. We evoke this recent memory of familial bonding when we visit the rotunda at the southeast end of our garden. The rotunda, reconstructed from the top of a McIntire-designed church that no longer stands in Salem, reminds us again of the fragility of the past, though it is framed by a majestic, long-standing specimen tree—whose roots are ironically now threatened by the new foundation of the bathhouse. This tree, an ancient ash, oversees our more recent endeavors in the flowerbeds. The tenuous maintenance of always-flaking treillage, and the tedious clipping of foliage in the garden rooms, depends on the men and women tending this garden, now and in the future. These garden spaces are dependent on our family interest in sustaining our claim to them. Will the animating current of recent memories be strong enough to reinforce our family connections, the flow of images that tells our story, and channel them into a growing sense of proprietorship? This is what our earlier ancestors must have felt when they cared for the ground of Greenwich Common.[27]

Such a claim to a place can, as the Greeks believed, depend on fate—the fickle intervention of the gods. We might interpret fate as the emotive power of DNA, a prime determinant of our interests and proclivities (along with family experience and interactions). This in turn affects the determinative power of family linkage and its capacity to sustain a legacy. Topophilia is a cultivated art, and many factors can deplete a family's energy and ability to care for a place. Trustees, fiduciary managers, and lawyers break up family properties; a family conservator passes away unexpectedly, leaving one hundred fifty years of family memorabilia to be scattered to auction houses; a family member assumes the care of an estate at nineteen, only to discover they are an entirely different person at thirty-eight.

We remain committed as a family to nourishing the sense of proprietorship we identified in our early ancestors on that town green in Greenwich. We seek to foster this energy in children who can identify with the alchemy of the *spiritus loci*—the spirit of a place. Our goal is remain rooted with our family tree, as represented in our garden. Other families have succeeded in continuing these "mystical chords of memory," as Lincoln once intoned. In great castles and mansions, there was a particular location assigned for the celebrating of ancestral connection, usually over a hall fireplace. For the Forbes family on their island fastness of Naushon, on the southwestern reach of Cape Cod, Massachusetts, this assigned place is a barn. We seek the comfort of such a framework at Bellevue House with our gathering point at our Family Chimney Tree and pizza oven, though the security of structure cannot mask the fragility of our individual fates.

The power of place still haunts and inspires me. The physical landscape of our garden is overlaid with a mental landscape, informed by a timeline that juxtaposes events in sometimes daring constellations. The spaces in the garden are populated with lost friends, glimpsed in the vistas of our Jeffersonian academical village with its arcaded stable block, itself the evocation of a lost Derby coach house near the Salem waterfront. The garden also contains private grief forever etched in my mind: the memory of a picturesque and tranquil Vietnamese village transformed by an American bomber into a shattered ruin. In its own curious way, the obstacles I encountered led me to garden making as a kind of therapy for my soul's torment. As my son-in-law, a devotee of Nietzsche's philosophy, might quote, "What doesn't kill you, makes you stronger." Far better for the garden to simply inspire, and for the power of place to sustain our family vision.

Notes

1 Marcel Proust, *Remembrance of Things Past*, trans. C. K. Scott Moncrieff and Terence Kilmartin, vol. 1, *Swann's Way* (London: Vintage Books, 1982). This concern appeared in my last volume: Ronald Lee Fleming, *The Art of Placemaking: Interpreting Community through Public Art and Urban Design* (London: Merrell, 2007). Proust's series of volumes, as rich as truffle chocolate, was the reading delight of our middle child. The two Adirondack chairs in the moss garden at Bellevue House are dedicated to our daughter's perusal of Proust the second time around. The Proustian idea of embedded memory influenced my own research for the placemaking books. One strategy for placemaking is to embed in the architecture and landscape objects (more tangible than a madeleine) that can conjure up memories in those who encounter them: for example, the bronze Californian landscape in the Smith Campus Center at Pomona College, or more abstract sculptural objects like the bronze sculls and VW bugs interposed with the volutes in the cascade garden at Bellevue House. I first articulated this concept in my early placemaker's book in 1981. We refined it with a better description of the technique of "environmental profiling" as a basis for examining the appropriateness of siting public art.

2 William H. Wilson, *The City Beautiful Movement: Creating the Northern American Landscape* (Baltimore and London: Johns Hopkins University Press, 1994). Wilson challenges the thesis that the City Beautiful movement arose from the exposure of about a quarter of the population to the great world's fair of the Chicago Exposition 1892–93. In fact, the first civic beauty association appears to be the Laurel Hill Association in Stockbridge, Massachusetts, of about 1854, and my book *On Common Ground* documents civic tree planting early in the nineteenth century in New England Green; see Ronald Lee Fleming and Lauri A. Halderman, *On Common Ground: Caring for Shared Land from Town Common to Urban Park* (Harvard, MA: Harvard Common Press; Cambridge, MA: Townscape Institute, 1982). Certainly, this enhancement instinct survives today with the enthusiasm generated by the Daffodillion, a million daffodils planted in Newport, Rhode Island, over the past nine years. They now mark the entries to the city at the bridge to the west, along the eastern and northern approaches to Middletown, with the newest planting (2019) at the seasonal park separating Middletown and Newport to the southwest.

3 The inscription is quoted from Fleming, *The Art of Placemaking*, 176–81.
4 E. W. Reynolds, a Connecticut Yankee and later merchant prince, took the train to Southern California in 1887 to recover his health. Beginning with a modest shop in Stamford, Connecticut, he later expanded to what became at one point the largest chain of wholesale stores in the West.
5 Holling Clancy Holling, *Seabird* (Boston: Houghton Mifflin, 1948); Holling Clancy Holling, *Tree in the Trail* (Boston: Houghton Mifflin, 1942).
6 Fleming, *The Art of Placemaking*.
7 My grandparents went on a round-the-world cruise on the P&O Line in 1922. The artifacts they brought back, which we discovered in file cabinets in the basement of their beach house at Hermosa, transformed my life at the age of four, as I have already recounted. The meticulously organized postcards from Japan to Warwick Castle in England gave me a vision, which the travel artifacts sustained.
8 Yi-Fu Tuan, *Topophilia: A Study of Environmental Perception, Attitudes, and Values* (New York: Columbia University Press, 1990).
9 I wrote the definition of *landmark* in the *Encyclopedia of the Environment*, where I included that phrase. See Ruth A. Eblen and William R. Eblen, eds., *The Encyclopedia of the Environment*, The René Dubos Center for Human Environments (New York: Houghton Mifflin, 1994), 397–98.
10 The library is a two-story Federal Revival–style wooden structure located in the garden of Bellevue House. It may be the only library in history that is combined with a nymphaeum; the challenge has been to balance the moist environment required by nymphs (water spirits) with the dry climate needed for storing books. The collections include roughly a thousand books on the history of cities and towns, which are organized alphabetically by place from Adelaide to Zanzibar. There are substantial entries on garden and architectural history, urban design and planning, historic preservation, cultural planning, and even a section donated by Caterine Milinaire Cushing on festivals.
11 Baroness Marjory Allen of Hurtwood (née Gill) was an Englishwoman my parents had never heard about—a landscape architect, a campaigner for children's welfare, and an influential activist for "adventure playgrounds." She was also the author of *Memoirs of an Uneducated Lady: Lady Allen of Hurtwood* (London: Thames & Hudson, 1975).
12 In fact, our suburb represented the conformity of the sanitized zoning of the 1940s—it was not an old-fashioned neighborhood with back alleys and a mixed residential area with poorer kids on side streets, as in Booth Tarkington's *Penrod and Sam* (New York: Buccaneer Books, 1975).
13 Jane Jacobs, *Vital Little Plans: The Short Works of Jane Jacobs*, ed. Nathan Storring and Samuel Zipp (New York: Random House, 2016); Jane Jacobs's essay "The Decline of Function" was published as "Do Not Segregate Pedestrians and Automobiles" in *Architect's Year Book 11: The Pedestrian in the City*, ed. David Lewis (London: Elek Books, 1965).
14 "We Europeans change within changeless cities, and our houses and neighborhoods outlive us; American cities change faster than their inhabitants do, and it is the inhabitants who outlive the cities ... to us a city is, above all, a past; for them it is mainly a future; what they like in the city is everything it has not become and everything it can be ... these cities that move at a rapid rate are not constructed in order to grow old, but move forward like modern armies; encircling the islands of resistance they are unable to destroy; the past does not manifest itself in them as it does in Europe through public monuments, but through survivals ... they are there simply because no one has taken the time to tear them down, and as a kind of indication of work to be done." Jean-Paul Sartre, *Literary and Philosophical Essays* (New York: Collier, 1962).
15 Smiley Residence, Canyon Crest Park, Redlands, California, postcard.
16 Albert K. Smiley, affectionately known as "Bert," has been president of Mohonk Mountain House since 1990. He is the great-grand-nephew of Mohonk's founder.
17 "Little Boxes," track 3 on Malvina Reynolds, *Malvina Sings the Truth*, Columbia, 1963, vinyl.
18 When our older daughter went to the college, I drove her luggage to her room in Claremont, then took her to the California Club, her great-great-grandfather's club in downtown Los Angeles, and let her take the train back out to Claremont as her grandparents had done in the 1920s and 1930s. The Spanish-style churrigueresque building is now landmarked and restored as the train station once again!
19 David Halpern, *Tulsa Art Deco* (Tulsa, OK: Tulsa Foundation for Architecture, 2001). Tulsa lost good buildings in the 1970s and 1980s, as their compensation outline of architecture reveals. To my knowledge, Oklahoma City did not produce a similar

inventory and appears to have suffered greater losses without the benefit of institutional memory.

20 This near miss occurred while I was escorting Larry Burrows as he photographed an ARVN (Army of the Republic of Vietnam) operation north of Saigon after the Tet Offensive; see chapter 4.

21 Deborah Wing Ray and Gloria P. Stewart, *Loyal to the Land: The History of a Greenwich, Connecticut, Family* (West Kennebunk, ME: Phoenix Pub, 1990). My ancestors, the Reynolds family, were among the founding proprietors of the green at Greenwich, Connecticut.

22 I can remember Thomas Boylston Adams's stories better than those of our own family. I recall him speaking about Augustus Saint-Gaudens's illustrious sculpture across from the Boston State House, which portrays Colonel Robert Gould Shaw leading his regiment of African American troops up Beacon Street to receive his appointment from Governor John Andrew, a family friend of Shaw's parents and also an abolitionist.

Adams told us, "My father was on the viewing stand for the dedication of the sculpture. Sitting next to him was an elderly lady, Mrs. Agassiz. She remarked to him, 'You know, it was usually quiet when the soldiers marched up the hill. There were no bands or cheering crowds. We were afraid that there would be draft riots, as there had been in New York, when the Irish were resisting being drafted for subscription in the Union Army. All you could hear was the slap of leather boots as they hit the pavement. It was so quiet that I could look down from our balcony and, as I had danced with him before he dropped out of Harvard, so I said, "Goodbye, Bob," and he heard me. He looked up and tipped his hat. It was the last time I would ever see him as he died leading those negroes at Fort Wagoner.'"

I still get goose bumps recalling that story, and I wish my grandfather had survived the trauma of the Depression, as he watched his fortune shrink, to tell us his stories. I wonder if he had really gone prospecting in the Klondike looking for gold before marrying Miss Mabel Reynolds of Los Angeles. Mabel Reynolds's own grandfather had been killed in the Civil War—we shall never know where else Benedict Ebner prospected. As Adams was fond of saying at his "soup and sole dining club" (which met at the Tavern Club)—a sort of inner sanctum of old Grotonians who understood Latin and could, when the occasion required, give elegant toasts—"What everyone knew then, no one knows today."

23 Christopher J. Koch, *The Years of Living Dangerously* (London: Penguin Books, 1983).

24 Jessie Sheeler, *Little Sparta: The Garden of Ian Hamilton Finlay*, new ed. (Edinburgh: Birlinn, 2015).

25 The lone carp, as brown as the surface of the pond, was our Darwinian specimen who grew fat wearing his camouflage, only to fall prey to an eventual oxygen shortage when workers changed the water level—a hazard of maintenance. I imagine he gasped in astonishment that his cover had been blown, like a fallen secret agent, and died, his sleek, corpulent body floating to the surface. Alas, there were no children around that day to give him a proper burial with a Book of Common Prayer. This rather poignantly demonstrates our need to install another generation of children at Bellevue House to invest the house with their own rituals.

26 The garden designer Gertrude Jekyll (1843–1932) and architect Edwin Luytens (1896–1944) collaborated on many Arts and Crafts gardens.

27 I haven't researched how long my ancestors maintained the right to keep cattle on Greenwich Common. On Boston Common, some proprietary rights apparently survived into the nineteenth century, I recall from my time as a trustee there. See Fleming and Halderman, *On Common Ground.*

Chapter Three

THE PLACE OF ANIMATION

Expressing the spirit

Just as important as finding the sense of place and embedding recollection in stored memory in a garden is expressing the spirit. This quality, which some may find nebulous, encourages animation—movement that can, over time, create its own "muscle memory" and contribute to the building of place meaning. This chapter is about creating a spirit that is animated so vividly that it resonates through time.

The garden can be consciously designed as an armature of memory that affirms the life of the family, and animating events can support the spirit of family life in a place. Rituals can evolve out of the stage that we set in the garden. The pattern language that creates the physicality of a garden (discussed in chapter 1) can combine with the rhythm of activities in the garden space to give it increased legibility in the minds of family members. We can expand this artistic armature when we invite the larger public world to enjoy the garden.

To drive memory forward, a narrative garden requires action, which the work of artists can embellish. Having a young family helped us to understand the importance of creating something that our family members would want to cherish because it would furnish their minds with memories of family activities and experiences. In the relatively new garden at Bellevue House, we would have to reinvent our traditions, build our family rituals afresh, as the childhood of our own children was rapidly vanishing when we acquired the house.

This garden needed to be a *tableau vivant* that could suddenly spring into motion. If it was going to be memorable in the lives of our family, the garden had to have the capacity to build new meanings that transcended passive contemplation of the pattern language. Our family and friends had to be actively engaged in the garden so that the accretion of actions in its space could build up their loyalty to it, their sense of proprietorship; for the young, this is best achieved through activities rather than word portraits about place and the vocabulary of its pattern language.

They had to dine in the garden, dance in it, barbeque and make pizza in it. They needed to invent their own celebrations with torch parades, afternoon teas, dedicatory christenings of follies, and dances in which graceful sprites wrap themselves in the limbs of the great beech, extend elegant forms across fountain basins, and dive neatly into limpid pools to choreographed rhythms. The garden had to be a place of merriment and good will in the lives of family, friends, neighbors, and, sometimes, supportive patrons for dinner and musical events. Ultimately, we had to build a sense of proprietorship that extended beyond ourselves to a broader constituency.[1]

When considering the Bellevue House garden as a stage set for action, it was appropriate that we made our most

Arts and Crafts Garden, Bellevue House (detail, fig. 157)

137

Fig. 137 Dancers at the emerald pool

Fig. 138 Dancers on the Villa Lante-style table

138

significant design interventions at gathering places. We recalled that artists, who became important architects of the Italian Renaissance, often began as set designers preparing for court fêtes, celebrating visiting potentates, organizing processions in the streets and masques in the piazzas. Their training was all about celebratory movement, usually in the baroque style that today we call *animation*.[2] They saw the possibility of quite literally "setting the stage" with dynamic design. The French social workers who enlivened drab neighborhoods with street performances after World War II were, in fact, called *animateurs*.

Family symbolism and ritual in the American Renaissance Water Garden

We designed our own garden in anticipation of the pleasure of people celebrating in its spaces. We sought to do this first in our American Renaissance Water Garden. The term *American Renaissance* refers to the period roughly between 1885 and 1930, when there were substantial efforts to integrate the work of artists and artisans into settings that told stories—allegories from the classics, mythology, legends, and poetry that celebrated the Renaissance concepts of the triumph of reason and man's command over nature. Such allegories were powerfully demonstrated at Villa Lante at Bagnaia in Italy, where wild gushing springs in the hillside above the garden are channeled into a trough to cool wine for celebrations at the famous garden table, which we replicated at a modest scale on our garden terrace, just east of Bellevue House and adjacent to the new breakfast room (figs. 137 and 138).

It was this sense of celebration that attracted us, but also the opportunity to work with the iconography, the symbolism that could be built into the design elements. We realized that one could reinforce the other, and that the act of feasting could be placed into a larger context where one became conscious of the role of feasting in the life of the family. We initiated this garden design with the image of feasting at a long table. I had first seen the image of the Villa Lante water table as a young schoolboy, and it stayed with me through the intervening years.[3]

Situated on one of the descending terraces of Villa Lante, the stone table is approximately seventy feet long, much longer than we could accommodate on our new terrace. The Villa Lante garden itself was considered by many, including Edith Wharton, to be one of the greatest Renaissance garden creations (fig. 139). Cardinal Gianfrancesco Gambara designed it at a time when cardinals were competing for Papal favor.[4] He conceived the long linear water table, with a central channel to carry water gushing from the powerful springs above and then consciously tamed into rills and sluices that rippled down to tranquil balustraded quatrefoil pools below (figs. 140 and 141).

In our American memory, growing up in Southern California's green agricultural garden, we were conscious that wild torrents in the Sierras and in distant river valleys were dammed up and then channeled into canals that traveled to the thirsty cities and farmlands of Southern California. We experienced nature's bounty in irrigated parks lined with rustic wooden picnic tables rather than in a grand estate filled with carved stone benches. As a family we recalled the picnic tables along a three-mile sward of tree-lined park corridor that divided the main concourse in Ontario, California, near Pomona College (fig. 142). As far back as the 1950s, families gathered there for 4th of July picnics, with tables organized by the Midwestern states of their origin. These memories joined with images of Villa Lante in a vision of our own family feasting together at a water table on our own terrace. From that vision came an

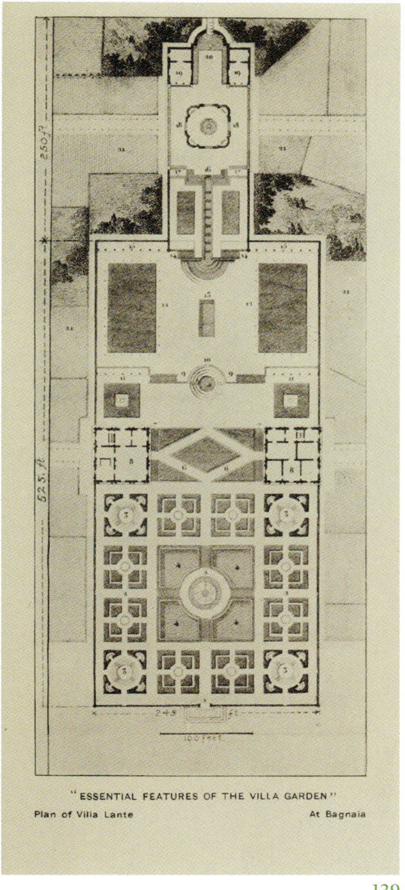

139

140

141

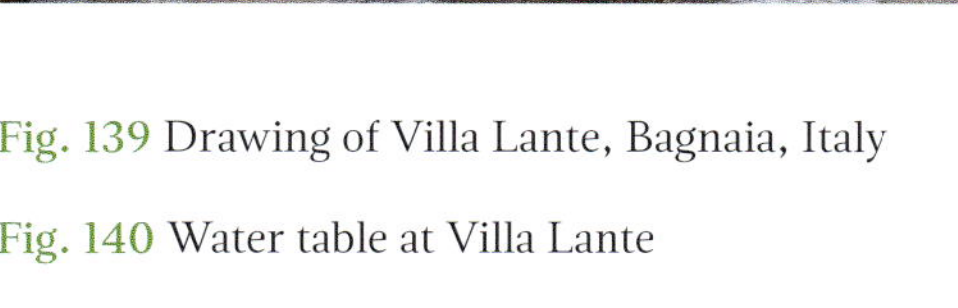

Fig. 139 Drawing of Villa Lante, Bagnaia, Italy

Fig. 140 Water table at Villa Lante

Fig. 141 Pools below the water table at Villa Lante

142

143

Fig. 142 Fourth of July picnic at Ontario, California, 1951. This was a down-home response to Renaissance frivolity, bringing the image forward to our time and place. The water flowing from the Owens Valley to the thirsty Southern Californian cities demonstrates the same mastery over nature expressed in Renaissance Italy.

Figs. 143 and 144 Goddess Pomona and the Villa Lante-style table in the American Renaissance Water Garden, Bellevue House

ordered sequence of design motifs that both created an armature for celebration and came to symbolize the formation and extension of family over time.

Our architectural design sequence begins with the goddess Pomona (figs. 143 and 144), the deity of the orchard, who stands with a cornucopia breaking open in her arms, her back to the lattice panels. She is carved to resemble my daughter, Siena. Poised in a fountain of gushing waters, she gazes upon the table and the waters at her feet, which splash down to the carved trough in the middle of the stone table (fig. 145). The table canal terminates in a head representing me, the garden proprietor, chiseled out of the stone mount at table end (fig. 146), facing the garden. The water descends through the mouth of the head into basins below and then streams under a stone footbridge and courses into a mosaic-lined channel below (fig. 147). The waters then traverse seventy feet down a narrow channel lined with granite lintels, which frame long panels of patterned, pebbled mosaics

144

145

146

Fig. 145 Villa Lante-style table

Fig. 146 Fountain head on the Villa Lante-style table

Fig. 147 American Renaissance Water Garden

Fig. 148 Rill in the American Renaissance Water Garden

147

148

that run through the narrow grassy garden enclosed by herbaceous borders. At the terminus, another carved stone with its own plashing jet completes the vista of the rill. It bubbles and splashes forward to the terminus of the garden, the elevated Children's Fountain chiseled from a single block of stone and embraced by two semicircular granite benches (fig. 148).

The metaphor, then, is that Pomona, the bountiful goddess of the orchard, spills her cornucopia of fruit and waters burst forth, nurturing the Family Fleming at their table. From the head of the father, the energy of life (water) flows into the basins and down the rill, representing the passage of life, until it reaches the children's fountain and curved benches that terminate this garden space. The three children, symbolized by monkey heads and wrapped in swaddling clothes (fig. 149)—imagery borrowed from the foundling hospital in Florence and from the Florentine artist Andrea della Robbia (fig. 150)—support the base of the fountain. Six circles are incised in the polished gray-green granite rim.[5]

Each Fleming child drew an image in stone in those circles that represents a core idea in their lives. Siena, an artist and with a recent doctorate in anthropology from the Sorbonne, created the most complex carving, a series of objects that symbolizes her life: paint brushes, a book, a loaf of bread, her heart, even a rib where a bone was removed to ease the passage of a life-threatening blood clot (fig. 151). Three circles remain empty so that in a future time, if the children decide to keep the garden, they can make new images of their vocations, their interests, and their dreams on the smooth stone or encourage their own children to create images and perhaps trace their own new sense of proprietorship. This narrative of family evolution continues into the Arts and Crafts Garden, now located directly on a continuing axis at the southern reach of the property, where the Arts and Crafts pool, which we also think of as the grandchildren's pool, extends the celebration of family life to the next generation (fig. 152).

A balustraded terrace with staircases overlooks the pool and frames a fountain tucked into the cove just below. In this sunken garden, which pays homage to the work of the English architect Edwin Lutyens, we can imagine the continuation of the energy flow from the American Renaissance Water Garden. Beneath the steps that lead down into the garden, to the large sapphire pool that is its central feature, is the figure of a dolphin, which represents my son, Reynolds Fleming, spurting water into a basin; the water then pours into the pool. In prerevolutionary France, *dauphin*, French for "dolphin," was the title given to the heir apparent to the throne. Here the family line is pouring into a pool representing family continuity.

The broken stone walls and pavements around our pool, the ceramic urns, and the plant pockets and millstones inserted in the stone walls recall the Arts and Crafts design vocabulary of Edwin Lutyens and Gertrude Jekyll's rill gardens at Hestercombe and by the Lily Pool at Buckhurst Park in England.[6] Lutyens also expressed a sense of animating ritual in some of his public works, such as the Rajpath in New Delhi, a vast ceremonial avenue flanked by government buildings and former maharaja palaces, which was designed to accommodate great inaugural processions. We reimagined, on a tiny scale, Lutyens's royal procession as an animation of family procession (and progression).

Just as the India Gate stands over the Rajpath (fig. 153), so our pool is marked by an arch, but one that relates in scale and character to the Federalist follies in the garden. This arch is modeled after the volute-bracketed arches that Samuel McIntire designed at the four entryways to the common in Salem, Massachusetts (fig. 154). Unlike the Salem arches with their relief of George Washington, the top of our arch depicts putti figures in bas relief pulling at Pomona's garments. The goddess is situated over the shower head inserted into the frame of the arch (fig. 155). She invites swimmers coming from the saltwater pool to stand under the shower and cleanse themselves—a ritual to celebrate new life as grandchildren frolic in a new landscape.

149

150

151

152

Fig. 149 Harold Peto-inspired bench and children's fountain at the end of the rill in the American Renaissance Water Garden, Bellevue House

Fig. 150 Tondo by Andrea della Robbia at the Ospedale degli Innocenti, Florence, Italy

Fig. 151 Drawing by Siena Fleming in the American Renaissance Water Garden, Bellevue House

Fig. 152 Arts and Crafts pool, Bellevue House

153

154

155

Fig. 153 India Gate designed by Edwin Lutyens on the Rajpath, New Delhi, India

Fig. 154 West Gate of Washington Square (Salem Common), c. 1850, daguerreotype, 10.795 × 14.288 cm (4⅝ × 4¾ in.), Gift of Mrs. William R. Cloutman, 1896, 3674

Fig. 155 Shower arch, Bellevue House

156

Fig. 156 Lotus fountain, Arts and Crafts garden, Bellevue House

Fig. 157 Arts and Crafts pool with lotus fountain in the foreground

157

A stack of eighteen lotus leaves, modeled after a fountain series that Lutyens created for the gardens of the Rashtrapati Bhavan (formerly the Viceregal Lodge) in New Delhi, rises out of the lily pond in the foreground and directs the eye to the southern rim of the pool (figs. 156 and 157). Symbolizing fertility, and the beauty of blossoms growing out of mud, the lotus leaves serve as the focal point of a newly conceived animating water ritual. I imagine monkey heads, one for each grandchild, that can be pulled on little rafts across the waters of the pool, from Reynolds's dolphin basin to the grandchildren's lotus fountain, to perch on the stacked lotus leaves. I envision a growing family of grandchildren lining the pool, cheering and pulling their younger siblings (as monkey heads) toward harmonious union with the lotus leaves. The infrastructure of the pool is conceived to embrace such animating rituals.

Ritual and meaning in the public spaces of the garden

Too many public buildings today are designed without ritual in mind, which in my judgment erodes the emotional content of the space they would create. The ritual can demand user-friendly symbols and define the way forward, but is made richer by anticipation. Rituals provide an armature of meaning with which we can invest public space. This can add to that muscle memory of animation. Many of today's modern buildings have ill-conceived behavioral patterns, which don't reflect how people use space, much less perform ritual.

For most visitors, interaction with the captivating images of our family evolution in the garden spaces will be through dance and music—festivals with lithe figures unraveling as they descend from trees, diving and writhing

158

159

160

in our fountains, or perhaps dancing in a great tent on the lawn, attending a premier musical performance around the moss garden, or listening to a brass quintet by the rotunda (figs. 158–60).

More recently visitors witnessed high school students playing Shakespearean roles in selected garden vistas (an innovation from the English-Speaking Union) and the Rhode Island ballet troupe performing on the south lawn, with dancers dressed like daffodils to celebrate the growing swards of yellow flowers lining the entries and shore lines of the city (figs. 161 and 162).[7] Inevitably, most visitors lay claim to the garden as a site of their own memory making, forming their own vivid associations with the garden, whether private and family focused or as a center point for civic animation, such as the public project to plant daffodils throughout Newport.

Figs. 158–60 Dancers at Bellevue House: at the emerald pool, in the beech tree, and on the common ground

161

Fig. 161 Dancers during the Daffodillion, Bellevue House

Fig. 162 Aerial view taken during the Daffodillion, Bellevue House

162

Creating a new common ground that can extend the spirit of animation

With white clapboard buildings constructed over a ten-year period, we incrementally shaped our own common ground at Bellevue House, and we began to see its unifying potential as a new community green (figs. 163–66). The last structure, the two-story stable block/garage, rose on the site of a dreary, crumbling green barn. This new architectural unity recalled to our mind another common space—one celebrated in my early book on greens and commons.[8] We saw Craftsbury Common, high up on a dominant hill in Vermont's Northeast Kingdom, as the archetype of these greens (fig. 167). Here was a carefully delineated rectangular space, crisply outlined in white railings set against the lush green grass of full summer. The common ground was lined with upright houses, and a church steeple rose above white clapboard; the architecture was sparsely ornamented, prim in its taut Yankee severity.

I first encountered this common in the summer of 1963 as a graduating senior from Pomona College and an obscure young Heritage Fellow from Historic Deerfield, visiting a professor and his wife from neighboring Scripps College. I was sitting behind two upright figures, stiff-backed New England ladies of a certain sort, motoring in an 1929 Pierce-Arrow automobile complete with running boards (fig. 168). The ladies were driving over from nearby Caspian Lake in Greensboro, a summer residence of academic families. They were bringing their own picnic to a strawberry supper on Craftsbury Common.

Our hostess that weekend, I later learned, was the model for the staunch lead figure in a Wallace Stegner novel that I read two decades later.[9] In retrospect, the coming together of these ladies and their friends of differing backgrounds for a strawberry supper was the singular event of that summer studying American history and the decorative arts. The ladies sat at a picnic table in the open air, the church meeting house looming above them. That modest, but poignant, event of country bounty and civility, welcoming in a stranger from Southern California, inspired me, a future narrative gardener. It was there that I affirmed a vision of America as a great and good place.

Fig. 163 Aerial view towards the west, with Newport Harbor in the distance

Fig. 164 Panoramic view of the common ground

Fig. 165 Library, stable block, and greenhouse

Fig. 166 Cabanas and cupola

163

164

165

166

167

168

At summer's pastoral end, I would be on my way to attend graduate school in Cambridge, Massachusetts, a long cultural distance from the brown hills of the California suburbs of Los Angeles where I grew up. Later, after years at Harvard, I went on to infantry school in the dusty red clay of a steamy Georgia summer, followed by a gloomy fall of industrial pollution in Dundalk, Maryland, training for intelligence work. I flew off to South Vietnam in time for the Tet Offensive. There, standing on a parapet on the edge of Saigon, I was narrowly missed by a Vietcong sniper's bullet, which whizzed past my right ear during a firefight.

Returning finally to Boston after a period of healing travel, I took a job as an urban analyst and later became an activist/preservationist running my own not-for-profit planning organization concerned with community visual character. A reporter at the *Boston Globe* interviewed me about my townscape design work, which was considered innovative at the time, and wanted to know about my experience pioneering main-street projects in scrappy New England towns. He knew I had been to Vietnam as a young officer. With a certain sardonic challenge in his voice, he asked, "What did you think you would be dying for over there?" I was a retired Army captain who had dined with twenty-four good men who did not return from this disillusioning war. I replied slowly, "If I had taken that sniper's bullet, I would have recalled a summer's evening at a strawberry supper on Craftsbury Common."

Now, more than fifty years later, as I contemplate the village I framed around the meadow space at Bellevue House, I realize that I have unwittingly created my own

169

170

Fig. 167 Craftsbury Common, Vermont

Fig. 168 The same 1929 Pierce-Arrow motor car I traveled to Craftsbury Common in, being driven by current family members

Figs. 169 and 170 Diodati Club gatherings in summer and winter, Adirondacks and Boston. Note former Secretary of State John Kerry on the right in fig. 170.

171

172

Craftsbury Common. My common is now intermittently occupied by community organizations: a dance company synchronizing those watery dives in the pools, a land trust, numerous garden clubs, hereditary societies, private social clubs, high school students performing scenes from Shakespearean plays, food and wine societies, daffodil lovers, beauty queen contestants, antique dealers, polo players, dance directors, associations of planners, and architectural historians (figs. 169 and 170).

All these people have experienced the subtle pleasures of this shared space, and amplified and animated its spirit. I believe that it is community benefit that sanctifies the common ground and that it is community sharing that kindles a sense of bonding there—a modest reaffirmation of a social contract now anchored in the collective entity of the groups that have shared the space. Consequently, the Rousseauian monkey on the tree trunk and his counterpart on the weathervane that peer across that space with their wooden and hammered-copper telescopes have, indeed, found a larger vision of a renewed social contract, supported by the humor of these crafted figures on the skyline.

As our handcrafted monkeys suggest and as the Johnny Nash pop song croons, "I can see clearly now."[10] This view can be applied to the evolution of our garden. It took a while for us to understand how the pieces could fit together, to comprehend how the whole could form something that was greater than the sum of the individual parts, to see that both townscape design and social fabric could be knit together again. The longer we lived with Bellevue House, the more we began to understand the collective pull of those parts.

I had, as a proprietor, spent a lifetime working on issues of townscape design, so I was already inculcated with the belief that the whole of a space should be more than the sum of its different parts. Up to this time, however (as recounted in chapter 2), I had built nothing larger than a ramshackle ghost town, so little buildings in a garden suited me (fig. 171). But putting them together and giving them a larger coherence was a much greater task. Orchestrating the animation of Bellevue House gardens to strengthen social cohesion in my community became my dream and my challenge (fig. 172).

Fig. 171 The "Hangman's Hotel" in the Fleming ghost town, Los Angeles County (note the boot hill markers). The Napoleonic figure is "Tom Sawyer" Fleming.

Fig. 172 A lifetime later, animating Bellevue House

Notes

1 I discussed this sense of proprietorship in my books on public spaces, building facades, and placemaking art, the focus of an earlier decade of work at the Townscape Institute: Ronald Lee Fleming, *Facade Stories: Changing Faces of Main Street Storefronts and How To Care for Them* (Cambridge, MA: Townscape Institute; New York: Hastings House, 1982); Ronald Lee Fleming and Renata von Tscharner, *Place Makers: Creating Public Art That Tells You Where You Are*, 2nd ed. (Boston: Harcourt Brace Jovanovich, 1987); Ronald Lee Fleming and Renata von Tscharner, *New Providence: A Changing Cityscape* (San Diego: Harcourt Brace Jovanovich, 1987); Ronald Lee Fleming, Rachel Goldsmith, and J. A. Chewning, *Saving Face: How Corporate Franchise Design Can Respect Community Identity*, 2nd ed. (Chicago: American Planning Association; Cambridge, MA: Townscape Institute, 2004).

2 Marella Agnelli, *Gardens of the Italian Villas: Drama and Delight* (New York: Rizzoli, 1987), 20–21.

3 A collector gave a huge trove of art history books to the private school I attended, Berkeley Hall, then in Beverly Hills, California. The books were stacked in a pool house and I would browse through them on my way to the swimming pool. Then one day they disappeared. The school had sold them to Pickwick Books in Hollywood rather than add their richness to the school libraries. I mourned the loss of this incidental source of knowledge—it was philistinic behavior, which schools are so capable of perpetrating—but I remembered the images of the Villa Lante water table in a book on Italian villas and gardens for the rest of my life.

4 Gianfranco Ruggieri, *Villa Lante*, English ed. (Florence: Bonechi-Edizioni, 1983).

5 John Pope Hennessy, *Luca Della Robbia* (London: Phaidon Press, 1980). Luca della Robbia (1399/1400–1482) was an Italian sculptor from Florence, noted for his terracotta roundels. He is remembered for the charm of his compositions rather than for the drama infusing the work of some contemporaries. Two of his famous sculpted works are *The Nativity* (c. 1460) and *Madonna and Child* (c. 1475). Sculptor Mark Mennin conceived the idea of integrating these images into our garden, demonstrating that appropriation remains a vital force in art making.

6 Edwin Lutyens, *Architectural Monographs* 6 (New York: Rizzoli International, 1979).

7 We hope visitors can associate this dance performance with Bellevue House's support for a million daffodils, a *daffodillion*, which we achieved in 2019 through the initiative of John Hirschboeck, our daffodil captain. This is thanks in part to the recent purchase of a Dutch bulb-planting machine, an innovation financed through fundraising at Daffodillion Dinners at Bellevue House. This allows us to plant 20,000 bulbs per hour. Before this innovation, the city claimed it only had the manpower to plant that many in a year. Now we can plant 200,000 a year, and have made our daffodillion.

8 Ronald Lee Fleming and Lauri. A Halderman, *On Common Ground: Caring for Shared Land from Town Common to Urban Park* (Harvard, MA: Harvard Common Press, 1982).

9 Wallace Stegner, *Crossing to Safety: A Novel* (London: Penguin Books, 1987).

10 "I Can See Clearly Now," track 7 on Johnny Nash, *I Can See Clearly Now*, Epic, 1972, vinyl.

Chapter Four

MEANING AND MEMORY COME TOGETHER IN THE ACADEMICAL VILLAGE

The northeast quadrant of the Bellevue House grounds is the site of our "academical village," a cluster of small buildings around a lawn, based on the one Thomas Jefferson developed at the University of Virginia. The village forms a distinct space in the garden, and its buildings, with their harmony of proportion and relationship of scale, shaped the garden's pattern language, which is expressed in a vocabulary of architectural elements including treillage, volutes, swags, and white wooden clapboard. Jefferson's academical village was a place designed to bring students and teachers together and foster collaboration; our village likewise hosts community organizations for conferences and other exchanges of knowledge. But the physical form of the village is also part of a larger landscape of the mind, a landscape of personal loss. In my mind's eye, the loggias of the village are populated by the ghostly figures of friends, the shadowy silhouettes and visages of lost companions who helped me see on my way. Their counsel shaped my sensibilities; their observations often structured my own powers of judgment; their aesthetic views sometimes reinforced a congruency, sometimes stimulated a reaction.

Fig. 173 and opposite Stable block and library/nymphaeum, Bellevue House

Lost companions: "Friends gone bye"

John Davidson Constable

Foremost among these lost companions is John Davidson Constable (fig. 174), who was a loyal friend and witness to several struggles in my life. With an English diffidence, he often bore hard truths with reserve, even taciturnity, though he also possessed a boyish enthusiasm that often discounted his reputation for Renaissance-man gravitas. He trained as a plastic surgeon and deployed his skills to help napalmed children in Vietnam, but he retained a modesty of bearing and even forgave the Vietnamese bureaucracy

173

for thwarting one of his healing trips to Vietnam. Here the combination of petite bourgeoisie French bureaucratic mentality combined with communist authoritarianism to block a border entry. Constable kept his patience and humility. He was a keen naturalist, having been a biology major in college, and he had a loyal following as an organizer of trips to exotic habitats for the Harvard Peabody Museum. He was even a devotee of the "speckled band" of Holmesian complexity who carefully reviewed the plot lines of Arthur Conan Doyle's works.

He was lean, angular, always nattily dressed, an actor in Tavern Club plays, a consummate club man. His sexuality gave him a certain vulnerability and empathy. We never openly discussed this, though he knew that I was aware of it. He witnessed key battles in my life: on the Cambridge Arts Council, on my board, and in the divorce struggle with my former wife. With that sense of English restraint and forbearance, he once said memorably, after an interview with my former wife, "There is evil here." But he was not one to say everything he knew, as many Americans are wont to do. He had an extraordinary sense of loyalty and was honored to help me manage my financial affairs.

He was connected to distinguished men and women around the world. He once told me to look up an antique dealer in Saigon who had helped him to form his great collection of Vietnamese ceramics (which he left to the Museum of Fine Arts, Boston); despite her bourgeois origins, she was a member of the Central Office for South Vietnam, the controlling apparatus of the Communist Party during the Indochinese civil war. He knew a Japanese finance minister; a Brazilian zoologist who owned a private island zoo; and members of Egyptian high society, which inspired envy in Egyptian doctors, who tried to impair his surgical skills in a mugging. He understood complexity, which probably fortified his compassion and nurtured his empathy. He is missed, but I see him clearly in the arcade of our coaching block.

Shot Warner

I first met Charles G. K. "Shot" Warner (fig. 174) when I was a trustee of the Society for the Preservation of New England Antiquities (now Historic New England). He came to me with a proposal that we obtain a National Endowment for the Humanities grant to archive the papers of Ogden Codman Jr., which were in Codman's family home, The Grange, in Lincoln, Massachusetts, where Shot also lived. He helped assemble the academics that must be fastened like ticks to any successful NEH grant application. We received the grant and, in the process, became friends. He was an usher at my wedding in Switzerland. Little did I know that, decades later, I would purchase Bellevue House in Newport, which Ogden designed for his third cousin!

Shot had a fuller life than many academics, and his skill as a raconteur was balanced against his insight as a trained historian. A member of the Fly Club at Harvard and a habitué of the ways of the Upper East Side of New York City, where his mother lived in an elegant townhouse just down the street from the Knickerbocker Club, he knew the gratin. But as one distinguished clubman marveled at his memorial service, he did not have a snobbish bone in his body. He was popular at Saint Paul's School in Concord, New Hampshire, where he gained his nickname for his skill as a sportsman, a pursuit he continued all of his life, as he attempted to teach me to cast a rod or find a duck (to no avail). With his brother Willy Warner—remembered for his 1976 Pulitzer Prize-winning book about the isolated island families of Maryland's Eastern shore, *Beautiful Swimmers*—he operated a ski lodge in Vermont after the war. He obtained his PhD at Columbia for his thesis on the vignerons of France, where he retained a share in a château in Bordeaux for many years. He doled out his remaining bottles to his friends, and I savored them to the end.

Shot dispersed his best stories to his friends rather than recording them for posterity. His account of seeing his tiara-sporting, powdered, and diamond-draped grandmother and her friend, Lady Decies, in a Weegee photograph airdropped by the Germans over the beaches at Anzio in Italy was a classic. The German propaganda leaflet on the beaches asked: "GI, is this who you are fighting for?" Shot tried furtively to gather them up. His epic liberation of Liechtenstein in an American Army Jeep and his adventures with artists in Paris should have made their way into a book! They were so droll and side-splittingly funny, but evaporated away in talk with his coterie of friends.

He taught himself Icelandic and read their sagas, studied astronomy, and collected French literature about Alfred Jarry's absurdist drama *Ubu Roi*. He introduced me to his boarding school classmates Henry Hope Reed (founder of Classical America) and Frederick Morgan. Morgan and his wife, Paula Deitz, were successive editors of *The Hudson Review*. Shot also connected me with some of his wife's kin who lived in the "Hudson River bracketed" world as Edith Wharton called it in her novel. He bequeathed me an English tailor and a French riding cape a size too small. I went back to Paris and purchased one in the right size in the only store that sells them; I wear it to this day in his memory. He also endowed me with a sense of the absurd, which has served as a useful antidote on occasion. He was a noted playwright at the Tavern Club and won the bruin on several occasions. In the end, he suffered from some form of Alzheimer's, but retained his disposition even as he lost his ability to keep his train of thought. He had many admirers, a humanistic complexity, and brought to every place where he hung that riding cape a certain bonhomie. He had a legendary charm that brightens an archway in my loggia of memory.

174

Fig. 174 Walkway from the stable block to the library, Bellevue House, with lost companions: (from left) Charles G. K. "Shot" Warner, John Constable, Lord Peregrine St. Germans, Pope Coleman, Lester Glenn "Ruff" Fant III, Rev. John D. Dennis

Peregrine St. Germans

Peregrine St. Germans (fig. 175) was a friend since my Wanderjahr after Vietnam, when I sought the healing repose of an English country house as a palliative to the shock and trauma of the war. Perhaps I am the only intelligence officer assigned to the Special Forces who made such a choice! Indeed, I may be the only American officer to attend the Attingham Park program of study on the stately home. This sojourn stretched into a year of travel that spanned the British Isles and extended to Europe and the Middle East.

Through a relative of Sir George Trevelyan, who ran the adult college that sponsored the Attingham study program, I was invited to Port Eliot for lunch and ended up staying for two weeks. Port Eliot was Peregrine's vast house in St. Germans, just south of Plymouth in Cornwall. Through his former brother-in-law Charles Shelbourne, now the Marquis of Lansdowne and a close friend of the Prince of Wales, Peregrine later sought to save Port Eliot as a training

Fig. 175 Lord and Lady St. Germans with Roo the whippet

175

center for rural employment, but through a miscalculation this did not come to pass.[1]

We bonded over a shared empathy. I admired his patronage of the arts, his capacity to go beyond the comfort zone of the landed Old Etonians of his youth, and his sense of adventure. It was a friendship that was to stretch over thirty-eight years, from 1968 to his death in July 2016. I came back to visit the first Christmas after I had left England, and Peregrine's father was in residence—he was called the "bookie Earl" of St. Germans. I recall his boredom with the life he had lived, as well as his interest in bedding the lady I had brought for the Christmas week. Though he asked permission, this also did not come to pass.

I repeatedly returned to Peregrine to mark key moments in our lives: he was an usher in my wedding in the cathedral of Bern, Switzerland; we visited each other in Cambridge, Hawaii, and London; and he even joined my mother and me for a meal at Maxim's in Paris. We communicated our joys and sorrows until various diseases crowded his later days. The obituaries spoke at his death of Byronic decadence, but I had witnessed a conservative side, a deep caring for the countryside where his family had lived for five hundred years. He and his best friend, Richard Carew Pole, drove me around quiet country lanes debating the angle of a house addition or roof line, opening my eyes to a refined sense of proprietorship. It was a profound teaching moment and reinforced our sense of empathy.

Pope Coleman

Edwin Pope Coleman (fig. 174) was an elegant scion of a Kentucky and Virginia clan who spent much of his life working for the public interest. I met him because he established the Hillside Trust, his vision for protecting the green slopes of Cincinnati from development. He spent much of his life in Cincinnati, where he fathered three children and was interim director of both the Contemporary Arts Center and School of Art at the University of Cincinnati. Coleman fought plans to build massive power lines from East Walnut Hills to Northern Kentucky, which would have crossed the Ohio River. The power lines eventually were buried underneath the river. He was a civic leader who inspired me with his capacity to give personal direction. He told me what to read and critiqued my leadership style and management capacity with grace and humor. A Harvard graduate, he was the son of a coal baron whom people called "the Count" because of his patronage of the arts in Lexington, Kentucky, where Pope grew up.

Pope was the handsomest of my friends, while still retaining modesty, a sort of Middle Border naivety,[2] but with elegant manners. He once drove me to the courthouse in Paris, Kentucky, so I could examine the records of my Kentucky Fleming ancestors who had settled in Bourbon County. I admired his capacity to intervene and his eleemosynary spirit. He retired early and lived in Santa Barbara and later in Carmel, which had a more liberal spirit. In the end, he and his wife, Constance, returned to Cincinnati, where they resumed their role in the community that had nourished them and that they continued to support, and where he had made a singular impact on the city. He was another usher at my wedding in Bern, and one of four who have departed.

Rufus Chase Stillman

My first hero was Rufus Chase Stillman, heir of a distinguished Connecticut family that had already shaped the character and legacy of the manufacturing city of Waterford. His ancestors had founded the Chase Brass Company and his great-aunt had established the Chase Ballet Company, and the family brought the majesty of an Italian city-state to Waterford's train station and city hall, expressing their pride of place. The Chases lived on the hill above the city, and I once spent a Christmas there.

Rufus had lost a leg at the Battle of the Bulge, but he had gone on to become a businessman-patron-collector who demonstrated his own passion and shaped the modern architecture of Litchfield, Connecticut. Rufus eventually led the Torin Corporation in nearby Torrington, where he and his own mentor, Andrew Gagarin—who also commissioned modern architecture and innovative design—worked. Rufus succeeded to the presidency but later lost the company in a hostile takeover.

I met Rufus's daughter Katherine at a lady's choice dance at Pomona near the end of my junior year. This turned out to be a momentous event. I told Kathy that I was to attend the Reserve Officers' Training Corps (ROTC) at Fort Lewis, Washington, near Tacoma, and that my mother was giving me a short trip to Europe afterwards. Kathy invited me to meet her family in the Loire Valley, where they were staying with the sculptor Alexander Calder and his wife Louisa. This was the beginning of a new friendship—with both Kathy and her parents, Rufus and Leslie.

My friendship with Rufus faltered for a while over the always-transparent fact that I never was romantically involved with Kathy, but rather maintained a platonic friendship. Very happily married himself, Rufus assumed later that my divorce must be my fault. Despite these complications, he was a father figure and counselor whose warmth of personality and social insight influenced my thinking during an off-and-on friendship that lasted through thirty-five years and three Breuer-designed houses (in the end, the Stillmans returned to their smaller first house). I eventually rejected the Stillmans' modernism and embraced the rich classicism of Bellevue House, which the Stillmans never saw. However, I admired Rufus's tenacity and his capacity to define the new architecture for Litchfield by doggedly sitting through school committee meetings to prevail on his alternatives to the traditional design of this beautiful and architecturally dignified town.

Rufus had a warmth that combined with his patrician style, and he would play bocce with the Italian workers in his factory. I admired this, and eventually came to see how he was influenced by these workers, and how intensely uncomfortable he could feel in a middle-class house in Litchfield when he took me to a meeting! I later understood his sense of shame that he couldn't protect his workers from the asset strippers who eventually broke up his company and sold off the pieces. Even when I saw him at his most vulnerable, I admired his honesty. I probably never felt a hurt greater than when he rejected me after my divorce, but he remained in my heart a great man. I regret that I learned about the death of his beloved wife Leslie in a roundabout way, and that I had been too marginalized to be able to comfort him in the letter I subsequently sent him.

Peter Boylston Adams

Peter Boylston Adams was the only son of Thomas Boylston Adams to fight in the Vietnam War—at the same time his father was running as a peace candidate for the U.S. Senate from Massachusetts. Peter attended Norwich Military Academy in Vermont. In many ways Peter best recalled the flinty, restrained character of the Adams clan, a family that held back its emotions. His father, Thomas Boylston Adams, set a tone of gravitas and probity.

Something of a juvenile delinquent in his youth, Peter became a banker after his Vietnam service and a family leader as a member of both the Somerset Club and the Tavern Club, where his acting in a play reminded us of his doughty versatility. His helicopters had been shot down five times in Vietnam, and he projected a toughness and self-discipline that impressed his academically inclined brothers, Henry, Douglas, and John, who all went to Harvard. Peter was the brother who braced up his cousin Murray Forbes and told him how he must behave in regards to his longtime girlfriend. He demonstrated a degree of old-fashioned New England rectitude that I admired.

He had been doused with Agent Orange by helicopter backwash in Vietnam, and eventually he received a disability status of ninety-five percent and moved to Florida. He and his wife Sherry would stop by Bellevue House on their way back from The Glades, the Adams family summer compound in a former sporting hotel on a peninsula stretching off Cohasset in Massachusetts. He had earlier recruited a carpenter-craftsman friend of his to serve as my caretaker, and continued to have a proprietary interest in Bellevue House and its slow regeneration.

Tightly constructed and of medium height, he had a certain intensity, a taciturn manner as compact as his frame. He didn't have the historian/moralist tone of his father, who wrote columns for the *Boston Globe* deploying tales of historic events as vehicles for moral lessons for our time. Yet in his own life, he displayed a crusty capacity to confront adversity. I never heard him complain. He possessed a kind of compressed gallantry, which included a certain amusement at the foibles of his more cerebral brothers. I miss his staunchness and his laconic grace. When he passed away, I raced to The Glades and arrived just in time to bear witness to his character at the memorial service.

Sergeant Wilson Fullbright

Sergeant Wilson Fullbright is another delayed casualty of the Vietnam War. Unlike Peter Boylston Adams, whose ancestors are enmeshed in the core of our national history, he remains largely unknown today. His parents are dead; his brother was killed in a traffic accident. His life of quiet service as a volunteer for the Preservation Society in Charleston, where he lived out his days in a wheelchair, a victim of a grenade thrown over a wall into the open courtyard of a restaurant, probably goes unrecognized.

To me it seemed his death was sudden. Maybe the exertion of having to take two hours to dress himself each morning, maybe an infection he never complained to me about, maybe the isolation of living alone in an apartment tower in downtown Charleston were all too much to bear. I sought to communicate with his attendant, but I never discovered the truth; in spite of a series of missed phone calls back and forth between Boston and Charleston, I simply knew that he was gone from our lives, if not our memory.

In searing contrast to the mind-numbing, self-cauterizing "Sorry about that" so often uttered in response to death in Vietnam, I was truly sorry to hear of Wilson's death, and I want to record this brief account to dignify his life before I too forget even this ghost of an outline, let alone the full impression of the beauty of his character.

Of course, I didn't know anything about Wilson's character back then in Vietnam. I didn't even know his name. I only knew that there had been a grenade thrown over the wall of Frigates, the best restaurant in Nha Trang, where I had once sat on a Saturday night, and that an intelligence officer had been badly injured.

Ten or so years later, after returning to the United States, I started my own planning office and did some pioneering work on main-street revitalization. I was in some demand as an urban critic, and one fine day I traveled to Charleston as a guest of the director of the Preservation Society. The director told me a story about their favorite volunteer and took me to meet him. He had been handicapped in Vietnam—*he* was the officer dining at Frigates, where the intelligence officers always met in the courtyard on Saturday nights, when a grenade thrown over the fence had paralyzed him. As we rode the elevator to Sergeant Fullbright's apartment, I realized that he was the man sitting where I could have been in Frigates that evening over a decade ago in Nha Trang.

Captain Ronald William Penn

Large in my landscape of personal loss is my dear friend Captain Ronald William Penn, the only Pomona College graduate in my year to make the Army his desired career path. He was an intense, dedicated, amusing officer who lived in nearby Cucamonga, California, only a few miles east of the campus. He was the son of an Okie who had successfully escaped the dust bowl and trekked west, like a Steinbeckian Joad family figure, and now owned three restaurants near the college. Ron was salutatorian of his high school class, and I was one of his two best men when he married Gladys, his high school sweetheart, and moved off campus his senior year. We were devoted friends, interested in military history, and fellow Republicans, his father representing sturdy self-made values, my Republicanism by inheritance dating back to the Civil War.

Though I doubled up on ROTC classes so I could escape for a semester and wear a beard (and be nicknamed Ron Flemingway), Ron Penn devoted himself to the martial arts and commanded Barrows Rifles, the ROTC unit's crack drill team, in his senior year. Later, when I was in graduate school, I visited Ron at the base in Germany where he was stationed after college. When I started my infantry training at Fort Benning, Captain Penn coincidentally showed up in the cafeteria line as an officer observer of our course. Over dinner at his quarters he said he might eventually command a country desk as an army intellectual. His wife, Gladys, was not enthusiastic, as they had adopted two daughters. I marveled at his dedication and enthusiasm for military life. He volunteered to endure the infamous "search and evade problem," which had us trying to avoid capture by sadistic sergeants who wanted to rough up green lieutenants in this overnight exercise through the red clay ravines of the piney woods. Ron loved it. I merely survived what was to me an exhausting ordeal—the roughest experience I had until the Tet Offensive in Saigon.

Later on, when I was still stateside in military classes, Ron wrote to me that he had finally gone to Vietnam. Gladys had put her foot down, and he had resigned his commission. When that happens, the Army has a contractual year left, and Ron had been ordered to Vietnam. But he had been given a job as a newspaper editor—a safe job, as he had a wife and two daughters and was leaving the Army. He wrote that although he had this newspaper job, he felt he must volunteer for a combat position if one became available. He reasoned that he had always been at the top of his class and had a responsibility to lead where his experience might save the lives of his men. He did not know about my own orders to Vietnam; I thought I would surprise him at the end of my tour with a stop at his base in Pleiku.

That journey took place after I arrived in the Fifth Special Forces Headquarters at Nha Trang and was assigned to the command bunker for intelligence. I realized I was going to be sent to a Special Forces team in the highlands with the Montagnard mercenaries, looking out through sand-bagged gun emplacements at an enemy force that was essentially holding us hostage. I thought there must be a better use of my talents.

I decided to invent my own mission to study the Central Office for South Vietnam (COSVN), the infrastructure of the forces that we were facing. I had the only book about it in Nha Trang. Fortunately, it had survived after my belongings were searched, though my Robert Indiana *Love* poster and a stash of anti-venereal disease drugs a doctor in Fayetteville had given me did not (the latter would have made me the drug king of the Special Forces). In the short hiatus between Colonel Kelley, who had confirmed that our mission was to "kill gooks," and Colonel Ladd, who had a PhD in political science, I seized my chance.

On a sleepy Sunday, with the Army operating under the order that the war would be won sooner if we took no day of rest, I entered the command headquarters where a captain was dozing at his desk. I had neatly typed the orders myself, which in effect said that First Lieutenant Fleming was authorized to use all available forms of military transport and housing and would be studying the infrastructure with visits to all four zones, called corps 1–4, from north to south. Lieutenant Fleming would report his findings back to the new colonel, who had yet to arrive. The dozing captain looked up and signed the orders, I left immediately on an adventure that took me all over South Vietnam.

The sheer audacity of this move for control of my own destiny still startles me to this day. It may have saved my life as well, since my interviews in Saigon with the Joint United States Public Affairs Office (JUSPAO) in the course of this six-week journey eventually resulted in my being asked to take over an information officer's post with the United States Information Service while he was on leave for three months. With the equivalency of a lieutenant colonel's rank, I was being asked to do meaningful work on propaganda cases of returnees to the government. I had never taken a course in psychology in college, but this didn't deter me from volunteering to take this job.

This eventual work would include the case of a young boy who the Vietcong propaganda machine said had blown himself up by detonating a mine against an American tank. There was a poster to prove this feat depicting Nguyen Van Bay as "the people's hero."[3]

Only he hadn't killed himself, and he was in our possession! We had concocted our own poster of him looking at the heroic image of himself. It was a rare victory for us, and apparently caused the Vietcong propaganda unit in charge to replace their cadre in three provinces.

The fact that another command wanted to hire me gave me a rare leverage I was to understand better later, but meanwhile I had delayed my trip to Pleiku in the highlands, only about an hour's flight away from Nha Trang to the west. I had saved the visit to my only friend "in country" until the very last before reporting back to the colonel. Alas, my surprise visit—as he didn't know I was also in Vietnam—proved to be too late.

I will never forget the day I arrived at Ron's base and asked a captain where I could locate him, only to be told I had arrived a week too late. Ron had volunteered to lead an armed reconnaissance unit: these units are, in fact, miniature armies charged with finding and sometimes engaging enemy forces, and perhaps the most dangerous assignment one could request. His unit was shelled in its first engagement. He was severely wounded and medevacked to Japan, where an army surgeon was unable to remove the steel fragments in his brain and he died on the operating table. I was in torment wondering if, had I arrived several weeks earlier, I could have reminded him that surviving combat is largely a matter of luck—indeed, that's what survivor's guilt is all about—and he had already served his country honorably and well.

I wanted to tell him what I had heard from then Captain Brian Jenkins, a colleague at the Fifth Special Forces headquarters in Nha Trang. In his first firefight as part of a Special Forces A-Team against the Montagnards in the highlands, the grizzled and battle-tested old sergeant next to him said, "You don't have to stand up now, lieutenant." I never had the chance to say that to Ron Penn. Later Jenkins, a painter in college, became probably the preeminent expert on terrorism at the Rand Corporation in Santa Monica, the think tank for the Pentagon. He now lives in Ojai, California, where his Vietnamese family cooked an elaborate multi-course dinner for my son and me one Christmas Eve. I felt destined to include this commentary as I hadn't seen him in fifty years.

Ron's wife Gladys had conceived a son while they were on leave in Hawaii, before Ron's fatal reassignment. I spent the afternoon writing her a letter of condolence in which I poured out my grief and sorrow. When I finally returned to California, I made no attempt to contact Gladys, as I wanted to remain true to my relationship with Ron. I suppose I was concerned that sometimes there can be an intimacy between a bereaved wife and comrades of her fallen husband. I had been second best man at their wedding.

Some years later I went back to Pomona for a rainy weekend class reunion, which had to move to the memorial gym because of the weather. I witnessed the plaques for the Pomona dead in all our wars; they did not include our single Vietnam casualty. My classmates were blithely circulating around the memorial wall, totally unaware of this omission. Later another classmate, Bill Claussen, who had served in Vietnam as an infantry officer, paid for a plaque, and the development office enabled me to speak about Captain Penn at the subsequent five-year reunion in defiance of the attitude of denial.

The mental landscape: A Vietnam viewshed

The search for linkage in the new townscape of the academical village set around the common at Bellevue House can be seen as part of a larger quest for community. The follies and various outbuildings, all of a piece but still individually articulated, recall my larger search for a community of friends and mentors who would add a collective sense of support. My garden also connects to larger narratives in other ways, evoking the physical landscapes of other garden places, such as the Katsura Imperial Villa and Studley Royal Park (see chapter 3), but also alluding to a larger landscape of the mind. The ghosts of my companions in the academical village are joined in the common ground between the rotunda and the coach house by a host of powerful images of the Vietnam War—signifying loss of innocence, idealism and its betrayal—which flare up with cobra-like intensity and haunt me to this day.

I still see in my dreams that glade at Fort Lewis, where young ROTC cadets spent part of the summer after their junior year, training to be officers. During a break in our summer of marching around in the forest, I remember sitting in a lovely fern-fringed glade singing that old Pete Seeger song, "Where Have All the Flowers Gone?" The song ends: "They've gone to graveyards every one. When will we ever learn? When will we ever learn?" Pete Seeger was the nephew of a celebrated poet, Alan Seeger, who had died in the First World War.

Pete Seeger visited our garden a short time before he died, at the invitation of his old friend impresario George Wein, who, with the support of Elaine Lorillard, had started the Newport Jazz Festival fifty years before. In my mind's eye I can see Pete, still spry at ninety-three, clambering up to the highest elevation point of the Bellevue House gardens, which we had designed to better see our own Studley Royal *tempietto*. This vision of Pete Seeger climbing the three-and-a-half-foot mountain in the garden has joined the vistas of garden history now part of my mental landscape. The innocence of that ferned glade sanctified with Seeger's lyrics will forever be juxtaposed against the pitted and cratered landscape of Vietnam.

Fifty or so years before Pete Seeger took in the view of my Newport garden, I received my orders to go to infantry school at Fort Benning, adjacent to Columbus, Georgia. The ostensible reason that intelligence officers were sent to infantry school was to help them communicate with the branch that they would be working with. But as the intelligence officers were, on the whole, a sophisticated lot and often from more affluent backgrounds, I speculated that it was also to humble them. I arrived early to look at the architectural restoration and coincidentally made some friends in this deeply conservative community.

I realized that I was viewed as both a privileged member of the establishment and as a Yankee from the north—I confess I didn't always understand the nuances of my hosts and their friends. But my Harvard degree attracted attention, and despite some gaffes both at the base and in the community, I found myself interviewed by Major

Fox, an intelligence officer at the Pentagon. He flew down to see me and intrigued me with his account of the Korea operations, where, he said, the Army did black operations with tunnels to the north and saw real action, whereas in Vietnam the CIA did the interesting work (an irony I would eventually understand when "the company" took me in after the Army felt threatened by my skepticism).

Major Fox didn't know, I suppose, that my infantry school instructors marked me as suspect. I had broken the first rule of survival: stay anonymous. I had laid out in my barracks room a copy of *Ramparts* magazine, an antiwar publication, which prominently featured on the cover a story about a Green Beret who had quit the Army. Of course, it also didn't help that I had left a rifle stacked on the field when summoned for a medical appointment: this was evidently a cardinal sin.

These faux pas would continue to pock-mark my "brilliant career."[4] So it came to pass that I alone received the assignment to go to Korea, while all about me young second lieutenants just out of college received orders that simply said "APO New York." They were not "sore afraid" in the biblical sense. They were clueless.

By this time, I had other problems at Fort Holabird, the industrial-smelling east end of Baltimore. I was at the time living on historic Bolton Hill in downtown Baltimore, in the townhouse of a doctor on sabbatical from Johns Hopkins. I had the ground floor with the Picasso over the fireplace; four other intelligence officers subsidized the rent on the floors above.

The counterintelligence classes had been so simplistic and slow paced that I took to reading Tolstoy's *Anna Karenina* in class. I probably should have been reading Dostoevsky's *The Idiot*. So I was at a low ebb when I went into the Pentagon with a "strack" (slang for a highly motivated young officer) who wanted to volunteer for Korea.

I remember the interview well: The officer—I suppose he was a major or a lieutenant colonel—looked up from his desk with annoyance. I announced that the young officer wanted to trade with me. The senior officer said, "We gave you Korea because we do important operations there, tunnels under the border, black operations, that sort of thing. Why do you want to trade?"

I had by now learned something in the Army: I didn't say that I thought following a generation of Radcliffe College women down to the city seemed a good place to further my skills. The officer followed up with, "Do you know what *APO New York* means?" I said, "No, I thought it meant New York." He looked at me with growing irritation. "It means Vietnam, lieutenant. How would you like to volunteer for Vietnam?" There was a little pause as I took this in. I then said, "I am morally ambivalent about our war in Vietnam, sir." He looked at me with what seemed to be a certain contempt. "Morally ambivalent, is that so? Where were you at school, lieutenant?" I said, "Most recently at graduate school at Harvard, sir." He looked at me with disgust, and with gritted teeth sputtered, "Harvard, morally ambivalent! You have a big chest, son. We are going to straighten you out. Not only are you going to Vietnam, but we are putting you in charge of a small detachment of intelligence officers and assigning you to the Fifth Special Forces Group!"

There must have been a stream of intelligence officers stationed at Fort Holabird who annoyed the Pentagon with order-change requests. After my time, they moved the intelligence school, lock, stock, and barrel, to the distant Fort Huachuca, an old Apache fighting post near Tombstone, Arizona. Despite several written requests, I have never received a copy of the list of intelligence officers killed in the Vietnam War.

Through the machinations of Barry Zorthian, a former editor of the *Yale Daily News* who went on to become the U.S. government press officer in Saigon, I found myself reassigned during my Vietnam tour. Zorthian enabled Colonel Lattimore—who had surreptitiously read a draft of my Christmas letter praising the work of American journalist David Halberstam (see appendix 2), a former *Harvard Crimson* editor who challenged the positive tone of reporting on U.S. actions in Vietnam—to have me reassigned to the Cambodian border. I think Lattimore had been suspicious of me ever since he had invited me to Thanksgiving dinner and suggested that I bring my friend Peter Kann, another former *Harvard Crimson* editor and then a cub reporter for the *Wall Street Journal*, where he later became publisher. When the colonel asked me to give a toast, I said, "Here is to Ho Chi Minh, who brought us here." He was not amused.

I protested my reassignment vigorously to Zorthian, who had advanced opportunistically over senior public servants to head JUSPAO (the Joint United States Public Affairs Office). I said I was doing serious work on a proposed radio program that, in my judgment, increased U.S. credibility by featuring a range of points of view, including a former Vietminh officer who was now an architect in Saigon. I was also working on discounting Vietcong propaganda films by using their own coverage. This had been the suggestion of the famous Colonel Lansdale, a distinguished antiterrorism expert, whom I had sought out for drinks in his elegant French colonial apartment in Saigon.

When the going gets tough, the tough throw a dinner party, so on my last night I invited about twenty people to my villa, which I shared with a United States Information Service officer, for Vietnamese food and American ice cream. I had invited some volunteers from VISTA (Volunteers in Service to America), a U.S. national service program, as well as a Yale graduate high up in the CIA—"the company," as we called it. He was the last to leave. As we strode down the driveway, he said he had learned so much from the VISTA volunteers, fluent Vietnamese speakers who, ironically, had protested the war while they were students at Yale and then had been sent to Vietnam. The company man put his arm around my shoulder and said, "We need to have a dinner party like this once a week." I told him that I was being sent to the Cambodian border the next morning. He said, "We will see about that." He had me driven out to Bien Hoa, a major U.S. air base known as "Pentagon East," in his own black Mercedes to see General Davis, a West Point-educated brigadier. The poor general had to offer me a cup of coffee and apologize; I was reassigned to work for the Central Intelligence Agency, "the Company," as we called it. I felt compassion for the general, whose only son had recently been killed. It was apparent to me that it was difficult to win against an insurgency, but easy to pull rank and beat each other up. Perhaps this apology was my only victory in the war.

In the final weeks of my time in Vietnam I experienced a dénouement, a kind of small-scale finale of folly. I was working for the CIA. I had been an irritant in the Army, having declared myself ambivalent about the war in Vietnam, and had been recruited by that more astute agency to work on a new staffing strategy for information officers. One day near the end of my tour, now in civilian clothes, I heard there had been an "accident."

I decided to fly out with the investigatory team. They were going to a village some distance northwest of Saigon, in enemy territory—"Indian Country," as American soldiers called it. The village was situated in a gentle landscape patterned with rice paddies and rhythmic whitewashed houses lined with bright magenta bougainvillea, nestled along the curve of a small and tranquil stream. I arrived to find the villagers in a state of shock, huddled in little groups, silent and blank faced, in the ruins of their houses. Our bombers had laid a stick of fire down the main street, killing approximately fifty-seven children and thirty-three adults.

What I witnessed was what I suppose we would call a "deliberate accident." Two Air Force colonels had used technology to deliberately target this location, which was a crossroads of Vietcong traffic. It just happened to be a village. It just happened to be a place full of people who had probably lived there for some generations. These villagers were just trying to survive the war; they probably knew that the Vietcong were meeting there, perhaps convalescing, maybe taxing the local population.

Fig. 176 Ron Fleming in Vietnam (far left, holding microphone), recording Vice President Hubert Humphrey

176

Like most Americans I spoke no Vietnamese, except for a few commands, and only fractured French. I was officially not there. I wasn't supposed to be there: I had heard there had been an accident, and had tagged along on a plane. Was I supposed to apologize? Could I say on behalf of our government that I was really sorry, so very sorry? No one was listening anyway. The villagers were mute, almost motionless, their lives transformed forever in a flashing instant. This is the other landscape that I carry around with me now. When I look out on my academical village, along with my little cluster of buildings I also see this Vietnamese village; the faces of the villagers are juxtaposed with the images of departed friends in the colonnades.

I remember now a saying that went something like "Be kind, as we all carry burdens." I have kind thoughts about those lost friends—"Friends Gone Bye," as I titled one of my photo albums. I also have regrets about our arrogance as a people. We Americans saw the Vietnamese as "gooks" over there; we depersonalized them. When we received our orientation briefing at the Special Forces camp in Nha Trang, just off the plane and hitting the blast of heat as we trudged our way across the melting tarmac, a colonel had asked, "Why are we here?" After a long, embarrassing pause, a young soldier responded, "To kill gooks." Apparently, that was the right answer. We were dismissed. The U.S. Congress has never ratified the international treaty that would allow American combatants to be charged with war crimes; even the officers charged with allowing American soldiers to shoot Vietnamese civilians at My Lai couldn't be prosecuted. This massacre was more provocative and personal than what I observed in the village in "Indian Country." Though the place name is gone from my memory, the image of our "deliberate accident" will scar my mind forever.

The Vietnam experience that shadows the landscape of my village green in Newport recalls that betrayal of idealism. Two photographs in particular encapsulate this. The first is a black-and-white picture of me tape-recording Vice President Hubert Humphrey at Vung Tau, a beach resort northeast of Saigon that was home to the National Training Center for Revolutionary Development, a training camp for returnees from the Vietcong side to the South Vietnamese side (Fig. 176). Humphrey had opposed the war when campaigning against President Johnson, but now, accompanied by the chortles of cynical journalists in the back, found himself comparing the returnee program favorably with the University of Minnesota, which he had attended. He said he didn't know about Harvard or Yale. He had become the president's lapdog.

The second image is one I took of Vietnamese peasant women, which recalls that peaceful village where I had witnessed the accidental/intentional bombing a few weeks before the Tet Offensive (Fig. 177). It also brings to mind my journey a few miles north of Saigon after Tet, when a line of peasant women averted their eyes as they filed by me, across the road from a small American outpost. At the time, I had moved into an apartment in a CIA compound where Peter Arnett found me while all hell broke loose, as the Vietcong advanced into our neighborhood. This scene is recalled in Arnett's book *Live from the Battlefield*.[5]

It was the actual battle scene Peter Arnett described that compressed the horror in quickly unraveling images. I had moved into the CIA compound the night before the Tet battle. The experienced CIA contract fighters immediately left and, with one other young CIA recruit, I remained to defend the missionaries also living there.

First I tried to kill a man in black pajamas in the adjacent building with a borrowed rifle—I have a lifetime of gratitude for missing this innocent "neighbor"—and then, in turn, missed being shot by a sniper. A diminutive GI, helmeted and sandbagged, called me up to a nearby apartment parapet. He pointed to a Vietcong position in a cemetery on a hillock a block away. I realized later that, at six feet tall, I was only there to draw the bullet that sucked air three inches from my skull. This is depicted in the *Years of Living Dangerously* cascade at Bellevue House.

Hours after this missed shot, I unwittingly stumbled into a South Vietnamese ARVN (Army of the Republic of Vietnam) interrogation shed. In dirty civilian clothes, I had no authority to countermand what I briefly glimpsed. It is the ultimate heart of darkness that scars my life.

A naked prisoner was hunched on the floor. He was "wired for sound": a field battery with attached wires was connected to his testicles; electric shocks were cranked up, supposedly forcing him to speak. I staggered out and never learned his eventual fate. Was he driven mad, and then executed?

Several days later, after the battle in my neighborhood, I was escorting Larry Burrows, the distinguished English-

177

Fig. 177 Vietnamese villagers

born photographer. We had bonded over photographs of his that I had seen in *Life* magazine before I had arrived in-country. He took a powerful sequence of pictures during a firefight, in which a crew member in a helicopter was killed in front of his eyes. His images caught the pilot weeping in a powerful indictment of the horror of war (fig. 178). Larry told me that he had worked in an English coal mine as a conscientious objector during World War II. He had the gumption to stick a camera in the face of a dying man; I admired this but never could do it.

He was busy photographing an ARVN unit off in a field on our side of the road. The North Vietnamese or Vietcong unit retreating after Tet was probably already in ambush position; the ARVN soldiers maneuvering in the field were probably deliberately avoiding them. The blank look of the passing women told me something was very wrong. I summoned Larry, and, as twilight approached, we prepared to drive my jeep back to Saigon. We stopped to say goodbye to the soldiers at the tiny American detachment sandbagged adjacent to the road. We shook hands with the lieutenant and a handsome, rangy red-haired sergeant who had returned from leave for just that night to say farewell to his buddies before going stateside. After we left, the Vietcong unit apparently zeroed in on the bright light of the television at the American camp. The popular sergeant was killed; the lieutenant was wounded in the throat. Not long after that, Larry died in a helicopter crash, on his way to observe the devastation we had wrought in Cambodia. The standard expression in Vietnam with which we attempted to immunize ourselves against pain or mask our indifference was "Sorry about that." But I never got used to it. I had shaken hands with twenty-four good men (one of whom was actor Errol Flynn's son, Sean) who did not come back; most of them, including Larry, I had met through my work with JUSPAO.

I survived my tour and the memories of the bombed and pitted landscape of that village in Vietnam. I went "back to the world," as we called it then, and not with the decompression time and camaraderie allowed by passage on a World War II troop ship. I was a single individual, dropped into a land of protests. I took a year off and studied the English country house to quiet my nerves—though, along the way, I also collected Czech political cartoons from the Prague Spring, was almost killed by a mob in a small village on a road to Casablanca, and survived an interrogation by an intelligence officer in Amman, Jordan, while the clock ticked on my Dutch student charter flight.[6] I had been escorted from my hotel to the cell block in the bottom of the central police station. Unknowingly, I had shown too much interest in the American girlfriend of what turned out to be the minister of the interior. I took a collegial approach to the intelligence officer who thought I was working for the CIA. I said if I was working I would have a cover, and proved I had some outside income by creating an ancestral tree on a black board. I made my flight.

When I finally returned stateside, I was recruited for a job evaluating President Johnson's Model Cities program. He was pouring money into distressed American neighborhoods. Neighborhood groups like the Black Panthers battled old-line city administrations for the funds. The funding would itself become a casualty of that war.

Returning to Harvard Square, I encountered riots there. As a result of my experiences, I proposed some changes at Harvard, and the then vice president for community affairs Charles Daley invited me to a cocktail party at his house. There I met David Halberstam. He had coined that expression "the best and the brightest," referring to the elite Harvard faculty brothers who advised President Kennedy.[7] Halberstam would later write the foreword to a book of Larry Burrows's photographs, which were posthumously published.[8] He was intrigued by my story of having become *persona non grata* on account of my support of his reportage.

I went to parties in New York, and Radcliffe College girls turned their backs on me. They now had a right to

178

Fig. 178 Helicopter crew chief James C. Farley (left) with jammed machine gun shouting to crew as wounded pilot James E. Magel (1940–1965) lies dying beside him, Vietnam, March 31, 1965

be rude, as I was labeled a war criminal. These were the same women I sought to join originally when, as a young intelligence officer, I had wanted the designation APO New York, rather than Korea (where the Army, not the CIA, was in charge and where I originally had been assigned before being sent to Vietnam). Agitated, I flew to Washington and enlisted in a peaceful march against the war. I remember someone behind me on the march saying, "I want to burn the flag"; I turned around and said, "If you do that, I will kill you." At this march I came across one of my Foreign Service officer friends, Frank Wisner Jr. He was, as he put it, just "observing from the sidelines." I had been a guest in his villa in the mountain resort town of Da Lat, where the Vietnamese on both sides kept a fragile truce so they could rest. I had also stayed with him and other Ivy League officers in an old Saigon villa that housed the last mistress of the last emperor of Vietnam. I remember being invited to the "Light at the End of the Tunnel" party, which included on its guest list top American officials such as Robert Komer. He had been an assistant to Frank Wisner when he was operations chief of the CIA and was now a guiding presence in the career of the charming Frank Wisner Jr.[9] The party mocked the optimism of a *Time* magazine editor who thought we were about to turn a corner and see the light.

A while later I attended a performance at the Hasty Pudding, a theater in Cambridge, Massachusetts. The play was titled *Medal of Honor Rag*: it was about a young, black soldier who survives a firefight in which his comrades are all killed, and he receives a medal.[10] Survivor's guilt causes him to hold up a store where the Polish-born owner (whose son was killed in Vietnam) is known to carry a gun. The soldier, though his pistol is unloaded, is promptly shot and killed. I broke down in my seat. Holding on with Waspish restraint, I cried after the audience left the theater, and only my girlfriend witnessed the agony of my grief. Shortly after this, I met the playwright, Tom Cole, at a local cleaner's and described my breakdown. He was gratified that his work had produced catharsis.

The ghosts of this journey from innocence in that fern-fringed glade at the ROTC camp at Fort Lewis to that scene of village devastation in Vietnam continue to haunt me.[11] They remind me how little our nation has learned from experience. I remember a phrase from my childhood: "Civilization is learning vicariously." But as Pete Seeger asks, "When will we ever learn?" My ghosts populate the academical village of Bellevue House and the Vietnam landscape is a permanent mental overlay on its common ground, a space where people come together to create meaning.

Notes

1 Horst and Valentine Lawford, *Vogue's Book of Gardens, Houses, People*, intro. by Diana Vreeland (New York: Viking Press, 1968).

2 This is a reference to characters in Hamlin Garland's Middle Border series of books, published between 1917 and 1928, which are well read and published to this day. They include *A Son of the Middle Border* (1917), *A Daughter of the Middle Border* (1921), which won a 1922 Pulitzer Prize for Biography, *Trail-Makers of the Middle Border* (1926), and *Back-Trailers from the Middle Border* (1928).

3 Don Rochlen, JUSPAO, Catalog of Psychological Operations Tapes, Tape 115.

4 The film *My Brilliant Career* was directed by Gill Armstrong in 1979 and was based on the novel of the same name by Miles Franklin.

5 Peter Arnett, *Live from the Battlefield: From Vietnam to Baghdad; 35 Years in the World's War Zones* (New York: Touchstone/Simon & Schuster, 1995) p. 247: "I felt I knew the city well enough to take off by myself, sometimes driving along side streets and alleyways to visit friends and acquaintances who were isolated by the fighting and fearful. A psywar operative I knew, Ron Fleming, let me through the fence of his Chi Lam home after I had leaned on my car horn for a few minutes. He cautioned me to be careful as we crept around the wall. Fleming's clothes were dirty and he hadn't shaved for a few days. He had a pistol strapped to his thigh and carried a high-powered M-14 rifle in his hands. 'There's no law out here, it's every man for himself,' he whispered, rising to his full height and aiming his rifle carefully over the wall and pulling the trigger. There was a Vietcong soldier hiding in the next building, 'the friendly neighborhood sniper' who had been harassing his home. The crack of the report echoed around the small cluster of neat concrete buildings in the compound. I was intrigued by Fleming's adeptness; he was a civilian who had graduated from Harvard University three years earlier, and now he walked around with the stance of an Indian fighter. As I was leaving, the neighborhood sniper had resumed shooting and Fleming was stalking him again."

While working at JUSPAO I sometimes attended the "Five O'Clock Follies." It was here that a memorable exchange between General Westmoreland and Peter Arnett took place, a perfect example of the cynicism of the press/government spokesmen: "Westmoreland's calendar for the years in Vietnam shows tennis games two or three times a week, with the preferred venue the Cercle Sportif, a club centrally located in Saigon. That created certain problems in terms of image and public relations. Westmoreland would show up at noon, just as, observed Robert Komer, all the reporters were having their breakfast. Eventually, harassed (in print) by Peter Arnett and Kelly Smith, Westmoreland gave up his membership in the club, angrily telling Arnett, 'You'll be happy to learn that I have quit the Cercle Sportif and tennis.' Arnett blurted out that, at his wife's urging, he himself had only recently joined the club. Then: 'Westy looked at me with narrowing eyes and said, "Well, maybe they gave you my slot," and turned away.'" Lewis Sorley, *Westmoreland: The General who lost Vietnam* (New York, Houghton Mifflin Harcourt, 2011), pp. 112–13. After the war I met General Westmoreland at a luncheon in Newport, RI. When he learned I was a former Intelligence Officer in Vietnam, he did not want to speak to me.

6 I made lifelong friends during my year in England, including Peregrine St. Germans, described above. Later, after his death, I commemorated him with his widow, then Lady St. Germans, in the *Years of Living Dangerously* cascade in my garden (see chapter 2).

7 David Halberstam, *The Best and the Brightest*, anniversary ed. (New York: Ballantine Books, 1993).

8 Larry Burrows, *Vietnam*, intro. by David Halberstam (New York: Knopf, 2002).

9 Wisner Jr. is disguised as the Foreign Service officer in Tobias Wolff's *In Pharoah's Army: Memories of the Lost War*, chapter 8, "Old China" (New York: Alfred A. Knopf, 1994). I still see him occasionally at the Knickerbocker Club; he now has a certain imperious manner that recalls Kim Philby's reduction of his father to one single sentence: "A short, self-important man, running to fat." Ben MacIntyre, *A Spy Among Friends: Kim Philby and the Great Betrayal* (New York: Crown, 2014).

10 Tom Cole, *Medal of Honor Rag: A Full Length Play in One Act* (New York: Samuel French, 1975).

11 I remain grateful that I was finally able to tell this story at a Tavern Club luncheon in Boston, through the good offices of Carol Bundy, herself the child of one of "the best and the brightest," in journalist David Halberstam's memorable phrase; see Halberstam, *The Best and the Brightest*. The Bundy brothers were part of President Kennedy's team that analyzed the military strategy that justified the escalation of the war in Vietnam. After the luncheon Carol Bundy returned to her house in Cambridge, where she broke down in tears.

Chapter Five

THE PLACE OF CRAFT AND THE ROLE OF THE ARTIST

Contemporary garden design has often minimized the place of craft. But some new gardens recognize the role of the artist. The New York Botanical Garden, for example, staged a show around Seattle glass artist Dale Chihuly, whose works dominated the garden's conservatories in the summer of 2017. In centuries past, the quality of garden ornament and sculpture, the setting of a special pavement, the shadow cast by an intricate iron gate on a garden path, or the magic of a shell-encrusted grotto could give a particularity to a garden design. Today, even well-endowed and popularly supported American gardens often have rather crudely executed detailing, which sadly demonstrates that the role of artists and artisans has been minimized in garden development. This, we believe, speaks to a lack of training of patrons and administrators in visual literacy, the "conscience of the eye."[1] Without such skills it can be difficult to recognize where key design elements are badly conceived or executed. In famous gardens like Descanso in Flintridge, California, we found an awkward, architecturally inept gazebo in a central location, and in Longwood Gardens in Pennsylvania we noticed clumsily designed footbridges.

However, there is no dearth of artists working today—we celebrated scores of artists, artisans, and their families at Bellevue House on May 4, 2019, as we neared completion of our garden. In our search for artists to commission for the garden, we found so much talent: artists working in glass, marble, wood, and bronze, crafting pieces small and large. Their work was more than competent and often superb; artists and artisans appeared available and as capable and competent as they had been in the nineteenth century, despite the complaints we sometimes heard about lost craft. We are more concerned about the artistic sensibilities of patrons and administrators who ignore these skills.

While we found enough artists and artisans (if not a plethora) who could meet our standards, garden design teams often do not include them or appreciate their larger role as problem solvers. Certainly, there does not seem to exist today the comfort level between artists and architects that arose when they trained together in the old *beaux arts* tradition. Stanford White and Augustus Saint-Gaudens were the best of friends and often traveled as well as collaborated together.[2] Today, unless a design competition or a "one percent for public art" law requires an artist's participation, when do we see an artist included in a design team? The good news is that there are more one-percent laws in the past few years and the most innovative and experienced public art programs, like those in King County, Washington, which includes Seattle, are using artists as part of art teams and on occasion as head of the teams.[3]

Villa Lante-style water table, Bellevue House (detail, fig. 180)

Cascading toward collaboration at Bellevue House

One of the ways that we learned about the necessity of artistic treatments was by viewing what happened to other garden places that failed to employ skilled artists—despite, of course, the best of intentions and adequate resources. One example is the facsimile of the Villa Lante water table in the recently conceived Chanticleer Garden in Wayne, Pennsylvania (figs. 179 and 180). In this case the machine-made simplicity of design, reinforced by the choice of highly polished black granite, has all the banality of a Formica lunch counter.

At Bellevue House we sought to remedy this problem of the generic solution that so often typifies garden design by engaging some thirty-five artists and artisans over a period of fifteen years. As a result, we have the *Years of Living Dangerously* cascade, with its explicit, personal symbolism; and in contrast to a generic weathervane, the monkey weathervane that sculptor William Reimann created for our garden (fig. 181): the commissioned piece has a complexity and a quality of workmanship that makes it unique. (It also took two years to wrest the monkey from Will; it came several times for cocktails only to return at the end of the day to the artist's studio.)

We believe collaboration with artists and artisans has added a level of complexity, energy, richness, and refinement to many of our design projects. However, some of that complexity also complicated the process and slowed down the schedule, as artists were not used to working within the tight budgets and deadlines of architect and construction teams. At least one outstanding artist in our collaboration had trouble understanding working drawings. Even determining the correct scale was a challenge for some. In one piece of ceramic work, we teamed two artists together: one had the figurative drawing skills to paint the tile and the other understood how to cast it. Several were

179

Fig. 179 Chanticleer Garden, Wayne, Pennsylvania

Fig. 180 Villa Lante-style water table, Bellevue House

180

perfectionists who had real trouble meeting a deadline. This can create costly delays as well as friction with contractors and the architect, and sometimes between artists.[4]

Sometimes a difference in interpretation can cause problems. In the many volutes we have deployed across our garden, there can be a tension between a loose artistic interpretation, where matching volutes bulge out of their architectural confines, and more precise work that supports a larger function, such as receiving a stream of water. For example, Mark Mennin's rather expansive volutes on the steps leading up to the Villa Lante-style table are deliberately carved to overreach its supporting step (fig. 182). At first, we were dismayed—and we know the architect was quietly incensed—but the overhanging bulge does have a certain edgy energy that compels our attention. We have come to admire it and think it gives the garden vitality. The larger, grander volutes that an anonymous artist carved (fig. 183) have a millimeter exactness to them: there is no room for expressive overbearance here.

In the *Years of Living Dangerously* cascade, the hollowed volutes that the architect designed to receive the dolphins' spurting jets had to be so precise that neither the monkeys on their backs nor passing wind currents could divert or deflect their stream of water. The volutes direct the water down to the central channel, where it must plash

181

Fig. 181 Monkey weathervane by William Reimann atop the cupola

Fig. 182 Overhanging volutes by Mark Mennin in the American Renaissance Water Garden

Fig. 183 Volutes by an anonymous artist in the *Years of Living Dangerously* cascade

182

183

into three bronze urns after passing the little sculptures set in patterned shell work divided by volute forms. The water is part of a sculptural artwork; each element must serve that larger context, that armature of meaning, and artist, designer, and artisan must collaborate to convey this in their work. To quote a colleague who worked on this book production, "Art is a conversation between order and chaos."[5] It is a visceral example of an ongoing tension between the exuberance of sculptural form and the mathematical constraints of proportion.

Monkey business modestly renews the social contract

The carved monkey representing a top-hatted Rousseau looking for a new social contract is another example of the symbolism that can be generated by engaging artists in a garden. The role of garden spaces in the social contract is something that has great meaning to me. For eight years, garden patrons have subsidized the planting of daffodils around Newport as a modest way of renewing that social contract. The daffodils, now more than a million strong, mark entryways and a golden necklace of key arterials around the city, as well as large swatches on the Salve Regina campus. As founders of this effort, though by no means the first to attempt such work, we looked for inspiration to the city of Inverness, Scotland, at the suggestion of Scott Wheeler, the city's tree warden. We initially thought to do the work anonymously through the Preservation Society of Newport County (PSNC), where I was a trustee; it seemed a good idea to sponsor work that went beyond their agenda. However, the PSNC apparently grew nervous that they would have to make an ongoing commitment and outed me.

I turned to the Alliance for a Livable Newport, and with the later addition of my colleague, John Hirschboeck, expanded our sights. John, a former advertising man, came up with the memorable slogan *Daffodillion.* He saw that we could expand beyond the thirty thousand daffodils that the city said taxed their planting limits. As the new captain of the Daffodillion, John inspected a bulb-planting machine in Holland that could plant a hundred thousand bulbs in an hour, and we increased our momentum. Raising the funds at a dinner party at Bellevue House, we began planting a couple of hundred thousand daffodils a year, and despite struggles with tidiness-obsessed maintenance crews, we have now met our goal and are approaching 1.2 million.

Along the way, we discovered that this modest civic investment can help renew the social contract between citizens and their government as well as give children and their parents the delight of sitting in the midst of daffodil fields (but please don't pick!). Now we are contemplating going off-island and contracting to support Department of Transport planting to line the most dramatic entrance to downtown Providence. We are also encouraging an entryway strategy for adjacent Middletown that supports infrastructure for underground wires. This finance strategy uses a Business Improvement District to create an assessment on property owners to pay for the work. Ultimately, the impacts may stretch beyond daffodils and encourage projects that upgrade urban infrastructure.

Encouraging policy change to foster more placemaking

Think about all of the generic pieces in most public parks: bike racks, tree guards, drinking fountains, gates, playground equipment—these are standard features and, usually, artists are not invited to bid on them. Inviting artists and artisans into public spaces can give them a special quality and, at best, some placemaking characteristics (figs. 184–86).

One strategy is to require that certain line items—such as pavements, gates, guard rails, lighting fixtures, bathroom tiles, bike racks, and tree guards—be allocated as commission projects for artists and artisans. This can reduce the purchasing of standard, generic street furniture and corporate commissioning of "plop art"—public art pieces that do not accord with their particular site.[6] This strategy can usually be better accomplished if there is an artist's brief: a thematic statement that provides a comprehensive view of the history and context of a proposed site, physical design parameters and constraints, and useful behavioral information about how the public use the space or site. Sometimes a brief also includes a list of materials used in local art traditions and resources that might respect the particular character of the area and support its pattern language.

This sort of environmental profile was not a requirement for many General Services Administration and National Endowment for the Arts public art projects in the United States, and the results sometimes stuck in the eye of the general public, which undermined future attempts to commission public art.[7] Critic Tom Wolfe observed that arts administrators were often trying to please five thousand people in the Manhattan art world.[8] An environmental profile and an artist's brief can reduce community resistance and make an artist accountable to a larger community,

184

Fig. 184 Decorative tree guard on Fifth Avenue, Louisville, Kentucky

Fig. 185 Palm tree bench in Coral Gables, Florida

Fig. 186 Elegant lamp post in Cleveland, Ohio

185

186

rather than just a few peers whose careers may profit from a splash of *épater le bourgeois*.[9]

Ensuring a percentage for art in gardening projects doesn't appear to have many precedents, though one does see the occasional crafted drinking fountain in a park setting, such as the trailside bubbler at Fresh Pond in Cambridge, Massachusetts, encouraged by the Cambridge Arts Council. At Pomona College Farm, an artist-designed welcoming gate and inviting ceramic-lined benches have been integrated into the landscape. In planning artistic collaborations for the Bellevue House garden, we initially sought inspiration from municipal projects that were either comprehensive in their integration in a single environment, such as a subway station in Stockholm, or comprehensive in their collective impact on place, like the artist-in-residence project in Cumbernauld, Scotland, a new town created in the 1950s as part of a plan to rehouse populations displaced by World War II.

187

188

189

Fig. 187 Drinking fountain at Fresh Pond, Cambridge, Massachusetts

Figs. 188 and 189 Pomona Gate at Pomona College Farm, California Ceramics by Mary Beierle

Fig. 190 The Family Chimney Tree wraps around the sides of the pizza oven

Fig. 191 Drawing showing details of the Family Chimney Tree

The Family Chimney Tree: Running ancestors to ground

When contemplating the Family Chimney Tree—a bronze relief portraying family chronology affixed to the chimney of the pizza oven that stands between the cabanas and rises on a stone platform above the classical pool at the eastern end of the garden (figs. 190 and 191)—I am reminded of what my late friend, Thomas Boylston Adams, a historian and a descendant of a family that gave the United States two presidents, used to muse over a small table at the Tavern Club in Boston with childhood friends from Groton School: "What everyone knew then, no one knows today." He was constantly thinking about the shift in our culture and the loss of memory that accompanied it, and wrote a popular series of articles in the *Boston Globe* seeking lessons from historical narratives that had relevance in today's world. When locating my own family narrative in the gardens at Bellevue House, a meditative work that has consumed much of my energy in the past twenty years, I had to contend with this condition of receding memory in my own immediate past.

I struggled to finish the Family Chimney Tree. I was compelled to reflect that my family, particularly my grandchildren, might be largely unaware of their provenance, even in this space that celebrates it. Here they could learn about ancestral associations, and a chronology that they had not had the opportunity to hear about on a grandfather's knee. We had started our family too late to be able to tell these stories to a host of grandchildren, and divorce had further dissipated the strength and fortitude that family association might recover. Perhaps it was too late to embed a memory in a wall and expect it to be implanted in a mind.

My Charlestonian friends, the Albert Simons III family, were able to achieve this sense of family connection even though they had moved to New York, where Albert practiced law. They maintained their place memory through family connections, fashioned like mollusks attached to an ancient pier. Their extended family remained intact in Charleston. The Simonses had married young and had two generations of children. Family memory was constantly rejuvenated in a tidal flow of ritualistic visits to their Charleston houses, where several generations of family could repeat their stories to younger members. Albert told me it was an art to reweave the ancient threads that bound them to place and to time.

I have visited aristocratic French families where, at twilight, against a backdrop of crenelated walls, I have heard similar stories of family resolution to affirm their place in time. They were less likely to accept the middle-class indulgence of divorce, and strove instead to find flexibility that allowed the larger claim of family linkage to prevail, whatever the discontinuity of genetic fate might reveal over time. In an age of mobility, it took deliberate effort to avoid dislocation and loss, to keep the power of place and family association intact and transcendent. It was a process that bred compassion, even empathy, for family members who might not know their genetic coding when young. Perhaps by osmosis, then, while sliding pizza out of our oven, our grandchildren will recover family memory. My own children have evidenced little interest in family history thus far.

Finding models for storytelling in public places

Finding inspirational models for integrating more comprehensive storytelling and defining the public will to achieve it remain one of our larger goals. The greatest influence on our own garden grotto was the integration of art and design in the Kungsträdgården subway station in downtown Stockholm (fig. 192). This is a space in which art is as important as engineering. The interior is carved out of rock in a series of chambers, complete with walls of amphorae of wine and architectural entablatures at the entry of the tunnel. Everything in the space reflects the history of the site above, once the location of a French garden and a palace.[10]

The Massachusetts Bay Transportation Authority's Arts on the Line program brought public art to subway stations in greater Boston in the 1970s and 1980s. As first chairman of the Cambridge Arts Council, I was able to secure half a percent for art on four subway stations on the Red Line running through Cambridge and Somerville, which the MBTA was renovating. Though not as daring and ambitious as the Stockholm work, the art project included *Glove Cycle* by Mags Harries, casts of bronzed gloves left on the escalator, which helped record the rhythm of movement in Porter station. James Tyler's *Ten Figures* featured life-size cement statues based on real residents of blue-collar Davis Square. Dimitri Hadzi's large abstract work *Omphalos* outside Harvard Square station did not fare so well: the MBTA was eventually forced to remove it as a safety hazard when a piece fell onto the transit island next to the old subway kiosk (converted to an international news stand, a much-loved landmark that closed in 2019).

192

Fig. 192 Kungsträdgården subway station, Stockholm, Sweden

Fig. 193 Bathroom tile at the Hult Center for the Performing Arts, Eugene, Oregon

In my work with the Cambridge Arts Council, I found the integration of the arts in utilitarian settings to be a fascinating challenge (fig. 193). We invited sculptor David Harding, town artist for Glenrothes in Scotland, to address the housing authority staff; as I recall, his appearance didn't transform the hearts and minds of the housing authority, but he did influence those of us who were trying to craft a strategy for Cambridge neighborhoods. We encouraged works for playgrounds, such as a giant rocking horse, and commissioned the integration of old map frescos in a library reading room and the insertion of a huge table and chairs in another school building, changing the scale and thus the experience of the space.

But the actual integration of art into a public setting like an underpass, school playground, or bus stop can be considered threatening; it goes beyond the precinct of curatorial oversight, usually centered around the art museum. An artist can find it challenging trying to create playground objects that meet safety and performance standards. The rocking horse in Cambridge, for example, was eventually removed from the playground because the elm wood it was made from deteriorated and young children attempting to ride it got splinters in their backsides. (Behavioral analysis might have protected those backsides.)

193

194

195

The Cambridge Arts Council's ordinance required that municipal departments spend a certain part of their budgets on public art, but some departments were not transparent and resisted us. We were particularly interested in insuring that ordinance would not foster "plop art," the sort of abstract sculpture that surrounded so many art museums across America. It seemed to us a sort of continuation of the single-purpose zoning that had characterized much of the early legislation encouraged by the *Euclid v. Ambler* ruling in Cleveland, Ohio, in 1926.[11] These laws sought to restrict land use to certain functions in particular areas as a way of protecting the health and safety of citizens. This early concern rather ironically ended up sanitizing our cities and decanting much of the liveliness out of the American cityscape. It also helped enforce the isolation of the "curatorial class" in cultural zones.

Jane Jacobs and William H. Whyte Jr.,[12] and other keen observers, have critiqued this loss of vitality, which extends to cultural districts as well. Art objects often appear clustered around the cultural buildings and thus divorced from everyday life. At times, art is often not even connected to the cultural institutions themselves! These zones are often marked by the alienated, free-floating sculptures of "plop art."

Some public universities have a better record of accountability on the issue of public art integration. The University of Oregon commissioned a planner to develop a strategy for integrating art into its new construction.[13] Progressive communities like Tempe, where Arizona State University is located, ran a competition for the design of bus stops, and King County, Washington, which formulated some of the earliest one-percent ordinances in the country, now have matured their programs to include arts teams, sometimes commissioning artists as leaders, to shape the ideas for a holistic strategy. Artists as team leaders may not create specific works of art but conceive of the larger environment that other artists may be commissioned later in the process to execute.[14]

Fig. 194 Grotto in the nymphaeum, Bellevue House

Fig. 195 Viewshed of a seascape by Tatyana Murray in the secret room of the grotto in the nymphaeum

Artist-shaped environments at Bellevue House

We wanted art in our garden, but not abstractions alienated from their context. We didn't view the art as a collection to be curated, but as an integral part of the narrative we were building. Certainly, the concept of an entire environment being shaped by the hand of artist and artisan influenced our own grotto design in the nymphaeum below the library at the Bellevue House.[15] Here the artist Tatyana Murray created a viewshed that lengthens the eye with layers of painted plexiglass (figs. 194 and 195) that match the tones of encrusted shells on the walls. A waterfall embraced by cast stone steps leads out of the grotto and reveals the connecting cascade, *The Years of Living Dangerously* (see chapter 2), a few yards to the south.

On a larger but still intimate scale, we found that one way to introduce the arts and crafts was to thematically tie the garden together with particular design elements that encouraged multiple expressions. We chose the depiction of monkeys to invigorate the garden design. *Singerie* was an important element in eighteenth-century decorating motifs in France and Germany. The monkey had a comedic role—literally aping the peculiar cultural aspirations of humankind. The French artist Christophe Huet and others depicted monkeys as musicians, artists, and men of letters (fig. 196).[16] With the arrival of Charles Darwin's *On the Origin of Species* in the nineteenth century, the theme appears to have been subdued (we speculate that perhaps the monkey was now seen as threatening rather than charming).

196

We have introduced more than a hundred images of monkeys in the garden and house, principally expressed in craft (figs. 197–200). Wrought in copper, carved from wood, chiseled in stone, depicted in paint and ceramics, and even captured in plexiglass in Tatyana Murray's layered view shed, the monkey reminds us now of our humanity, with all our fragility and foibles. They are also a babysitting device and can be relied on to occupy the attention of children with hours of counting (to the relief of harassed parents who just want to view the gardens in peace or take a nap after a tiresome ride to Newport with children scuffling in the back seat of the family car).

One of the particular joys of the garden artworks was working at a variety of scales. We found great pleasure in fashioning modest and tiny details that encourage close inspection. One example is the mortise joints of the stone lintels that frame the mosaic panels along the rill in the American Renaissance Water Garden and recall to mind Achille Duchêne's mortise-and-tenon design for a lintel-framed planting bed in the National Museum of Decorative Arts in Buenos Aires (fig. 201), once a grand private mansion. Instead of using tenons, however, we connected our lintels with stone pegs (fig. 202).

The cast-iron trellises in the yew and hornbeam allée provided us with another opportunity for classical intervention. The iron uprights of the trellis are terminated with tiny bronze capitals (fig. 203) and have entablatured bronze feet (fig. 204), the work of Jonathan Glatt. Similarly, the William Chambers-style bridge in the Oriental Vale

Fig. 196 Painting of monkeys by Christophe Huet, postcard

197

198

199

200

Fig. 197 Billiards-playing monkey carved from a dead beech tree

Fig. 198 Monkey head on the Villa Lante-style table

Fig. 199 Flute-playing monkey

Fig. 200 Swaddled monkey in the American Renaissance Water Garden

has a railing mounted with miniature metal temple bells (fig. 205), as depicted festooning Chippendale chinoiserie beds in Chambers's design drawings. (We employ them as fittings for lights on the often-slippery curve of the wooden bridge.) Our obsession with detail benefits from the acute eye of our architect-designer and collaborator J. P. Couture. These tiny touches support the postmodernist maxim "Less is a bore" that is often used in much larger contexts. But there is pleasure in surprising, even confounding, visitors with these supposedly incidental but in fact very deliberate encounters.

This notion of subtle interventions that won't be seen at first or second glance is further expressed in other examples. A marble drain (fig. 206) recalls a design detail of a sofa carved in wood by Samuel McIntire (fig. 207). The drain is located under the inlaid marble table we commissioned in Agra, India, for the greenhouse (fig. 208).

201

202

Fig. 201 Achille Duchêne's garden lintels with mortise-and-tenon joints at the National Museum of Decorative Arts, Buenos Aires, Argentina

Fig. 202 Lintels joined with pegs in the American Renaissance Water Garden, Bellevue House

203

204

205

Fig. 203 Bronze trellis capitals

Fig. 204 Bronze trellis feet

Fig. 205 Metal temple bells on the Chippendale bridge in the Oriental Vale

206

207

208

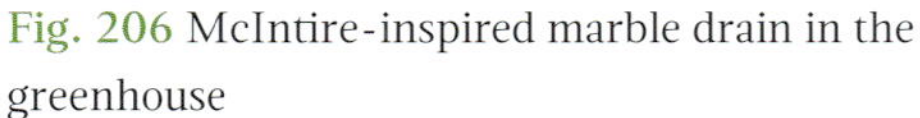

Fig. 206 McIntire-inspired marble drain in the greenhouse

Fig. 207 Detail of carved sofa by Samuel McIntire, 1799–1805, Museum of Fine Arts, Boston, The M. and M. Karolik Collection of Eighteenth-Century American Arts, 23.21

Fig. 208 Inlaid stone table in the greenhouse

209

210

211

212

213

Fig. 209 Mosaic in the Kitchen Garden

Fig. 210 Drain in the Kitchen Garden

Fig. 211 Front door

Fig. 212 Detail of sidelight next to the front door

Fig. 213 Detail of rotunda stairs

214

215

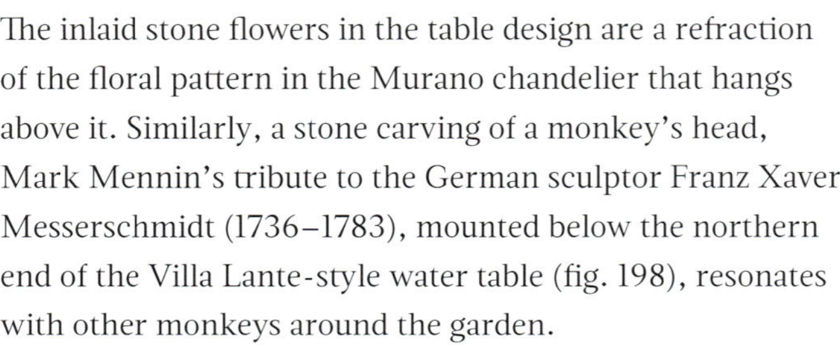

The inlaid stone flowers in the table design are a refraction of the floral pattern in the Murano chandelier that hangs above it. Similarly, a stone carving of a monkey's head, Mark Mennin's tribute to the German sculptor Franz Xaver Messerschmidt (1736–1783), mounted below the northern end of the Villa Lante-style water table (fig. 198), resonates with other monkeys around the garden.

Other references percolate through the house and garden. The carving of the drains in the kitchen garden echoes the pattern of a mosaic and the side lights on the front door (figs. 209–12), which in turn are expressed in the baluster design of the sweeping staircase in the rotunda (fig. 213), the most glorious feature of the house. The image of the goddess Pomona with the Putti disrobing her (figs. 214 and 215) is a sly allusion to the McIntire-inspired "shower gate," which is in turn a reference to the larger-scale gates that the city of Salem commissioned from McIntire for the four corners of Washington Square. The city replaced the arches at the time of Salem's bicentennial and failed to replicate the original design (fig. 216); we copied the McIntire design (fig. 217) only to discover an old photograph of the more elegantly carved and bracketed originals (fig. 218), which we then reproduced (fig. 219).

Bringing art and craft into the garden has allowed us to enrich and expand the complexity of the existing details of Bellevue House itself (see fig. 220) and bring them into the pattern language of the garden. It also amplifies that pattern language with new details and extends them through the property.

Fig. 214 Pomona's garments on the shower arch, Bellevue House

Fig. 215 Putti pulling Pomona's clothing on the shower arch

Fig. 216 Bicentennial renovation flattening McIntire's original arch in Washington Square, Salem, Massachusetts

Fig. 217 Reproduction of the flattened arch, Bellevue House

Fig. 218 West Gate of Washington Square (Salem Common), c. 1850, daguerreotype, 10.79 × 14.28 cm ($4\frac{5}{8} \times 4\frac{3}{4}$ in.), Peabody Essex Museum. Gift of Mrs. William R. Cloutman, 1896, 3674

Fig. 219 Corrected shower arch, Bellevue House

216

217

218

219

220

Fig. 220 The Library is influenced by the arts and crafts interiors found in Craigside and Arundel Castles, both constructed in the 19th Century with overhanging balconies that introduce a sense of snugness. The balcony railing motif at Alnwick, the Northumberland home of the Percy Family, particularly influences our design. This space was a collaboration between JP Couture and Ronald Lee Fleming.

Notes

1 Richard Sennet, *The Conscience of the Eye: The Design and Social Life of Cities* (London: Faber & Faber, 1991).

2 Paul R. Baker, *Stanny: The Gilded Life of Stanford White* (New York: Free Press, 1989).

3 In 2005, 53 programs incorporated 1% for art in private development programs, an increase from 39 in 2003, many in California. Currently the percentage generally ranges from 0.5–2% percent in the United States, and California continues to lead the country. The percent-for-art programs administered in Los Angeles make it overall the largest initiative in the United States. On the East Coast, the classic 1% is favored by a number of cities, including New York. Art is mandated in state-financed buildings through art-enrichment programs in 25 states. On the federal level, the percent-for-art initiative is administered by the General Services Administration Art in Architecture Program. The program earmarks 0.5% of the estimated construction cost of each new federal building to commission project artists. Overall, 46% of public art programs have a percent-for-art ordinance. Public programs are more likely than private programs to receive funding from a percent-for-art ordinance (66% and 14%, respectively). Sixty-seven percent of programs serving areas of 1 million or more receive funding from a percent-for-art ordinance or policy. Additionally, 48% of programs with a population between 100,000 and 999,999 and 20 percent of programs with a population under 100,000 receive funding from a percent-for-art ordinance or policy. See Sheppard Mullin, "Public Art Programs: 1% for the 99% – Part One," *Art Law Blog*, Art Law Gallery, October 10, 2012, https://www.artlawgallery.com/2012/10/articles/artists/public-art-programs-1-for-the-99-part-one/. See also Ronald Lee Fleming, *The Art of Placemaking, Interpreting Community through Public Art and Urban Design* (London: Merrell, 2007), 299–307; Americans for the Arts, *2017 Survey of Public Art Programs* (Washington, DC: Americans for the Arts, 2018), https://www.americansforthearts.org/by-program/reports-and-data/legislation-policy/naappd/2017-survey-of-public-art-programs.

4 When artist John La Farge didn't complete his commission in time for a Vanderbilt opening party, they retaliated with calls to their friends who had also contracted his services, persuading enough of them to cancel their contracts that La Farge lost so much business he had to declare bankruptcy. He went off to the South Seas in a funk with Henry Adams, whose wife had committed suicide. This doesn't seem a productive remedy for breaking a deadline.

5 Richard Sauvé, correspondence to a friend, shared with the author, 2019.

6 Alyssum Skjeie, "James Wines: The Architect Who Turned Buildings into Art," *Storyboard* (online journal of the Carnegie

Museum of Art), July 8, 2015, https://storyboard.cmoa.org/2015/07/james-wines-the-architect-who-turned-buildings-into-art/.

7 Fleming, *The Art of Placemaking*, 296–97, 317–20, 315–16.

8 Tom Wolfe, *From Bauhaus to Our House* (New York: Picador, 1981).

9 I document this approach in my placemaking books: Fleming, *The Art of Placemaking*; Ronald Lee Fleming and Renata von Tscharner, *Place Makers: Creating Public Art That Tells You Where You Are*, 2nd ed. (Boston: Harcourt Brace Jovanovich, 1987); Ronald Lee Fleming and Renata von Tscharner, *New Providence: A Changing Cityscape* (San Diego: Harcourt Brace Jovanovich, 1987).

10 "Kungsträdgården," The Art of the Subway, Visit Stockholm, https://www.visitstockholm.com/art-in-the-subway/kungstradgarden/.

11 Village of Euclid v. Ambler Realty Co., 272 U.S. 365 (1926), https://supreme.justia.com/cases/federal/us/272/365/case.html.

12 William H. Whyte Jr. was my mentor. "Holly," as his colleagues called him, did pioneering camera studies of how pedestrians used public space. He was visiting me at the time of the last redesign of the Copley Square plaza in Boston, which the Townscape Institute satirized in our *New Providence* poster series. We showed a sunken plaza so obscured from the street that a holdup was taking place in broad daylight. Holly was the behavioral consultant for the new design competition, and he was under pressure to influence the decision to go to a Harvard-connected landscape firm. He was not in favor of this firm's proposal, as it did not recognize the diagonal desire line that pedestrians would normally use to navigate the square from east to west; I encouraged him to stand his ground. This refusal to acknowledge common sense about how people actually used public space is a flaw in many modernist design proposals. See Albert LaFarge and Paul Golderberger, *The Essential William H. Whyte* (New York: Fordham University Press, 2000); Anthony Flynt, *Wrestling with Moses: How Jane Jacobs Took on New York's Master Builder and Transformed the American City* (New York: Random House, 2009).

13 Jane Jacobs, *Vital Little Plans: The Short Works of Jane Jacobs*, ed. Nathan Storring and Samuel Zipp (New York: Random House, 2016); Jane Jacobs's essay "The Decline of Function" was published as "Do Not Segregate Pedestrians and Automobiles" in *Architect's Year Book 11: The Pedestrian in the City*, ed. David Lewis (London: Elek Books, 1965). I once listened to Jane Jacobs lecture at the University of Colorado in 1969–70. She was a short, mousey bespectacled figure full of trenchant analysis, who took a critical view of the often antiurban behavior of planners, architects, and administrators who didn't understand the dynamics of urbanism. While I never heard her speak directly on precincts where cultural institutions are aggregated together—where they can't infect the rest of the community—her dry talk led to this conclusion.

Certainly, the instincts of David Oxtoby, the late president of my own college, and his art museum director, Katherine Howe, expressed this urban naiveté. They viewed the new museum building as a "border" (Oxtoby's word), rather than as an entryway through which people in adjacent buildings in this village setting would be integrated into the cultural life of the college and vice versa, as in the early college days, when Pomona College had a common that included the community. Instead, the trustee supervising the project spoke of the "collection" that they would be plopping outside the new building rather than of the art that would be integrated into the building. Despite a citizen lawsuit, in which the neighborhood was legally outgunned but made salient points of environmental impact, the college made only a token commitment to the community. Rather than programing community space and seeing the rooms as a way to bring prominent curators and collectors into the college, the college just powered forward, presumably so the president could claim the credit for an art museum. Pomona appears not to have seriously courted collectors in recent years as the meager collection attests, despite the wealth of arts riches in Southern California.

14 Fleming, *The Art of Placemaking*.

15 Barbara Baert, *Nymph: Motif, Phantom, Affect; A Contribution to the Study of Aby Warburg (1866–1929)* (Leuven: Peeters, 2014).

16 *Singerie* is a genre typically depicting monkeys mimicking human behavior, often fashionably dressed and with a touch of satire. Singeries were very popular as wall decorations during the French Rococo period. Although most say that French artist Christophe Huet's rooms, La Grande Singerie and the La Petite Singerie, created in the eighteenth century in the Château de Chantilly, are the most famous examples of singeries, monkey art has been around since the late Eighteenth Dynasty of Egypt and its popularity has continued into modern times. See Lynn Byrne, "Singerie: French for 'Monkey Tricks,'" *Décor Arts Now: Art in the Everyday*, November 29, 2014; Nicole Garnier-Pelle, Anne Forray-Carlier, and Marie-Christine Anselm, *The Monkeys of Christophe Huet: Singeries in French Decorative Arts* (Los Angeles: J. Paul Getty Museum, 2011).

Chapter Six

THE PLACE OF HUMOR

Narrating stories in garden spaces can introduce broad humor or its subtler and more refined companion, satire. Certainly, the manipulation of space can also introduce a playfulness, tricking the eye with juxtaposition of scale and prospect. The use of historical references in building design within the garden can challenge viewers to ask what time they are in—past, present, or maybe past-present. There is a serious integration of harmonies in building proportion, but juxtaposition can create a certain edgy tension. This edginess may be a deliberate provocation. Maybe we shouldn't take ourselves that seriously in a garden of cool harmonies and vistas. Humor tends to introduce a level of self-consciousness, anchoring us to the here and now. This can be subversive when the main intention of a garden may be to transport the visitor into a sense of the sublime. A belly laugh can distract from the larger vision, but maybe there is a place for laughter while contemplating the sublime.

There are some precedents for humor and its bedfellow satire in older gardens. At Hellbrunn Palace near Salzburg, the outdoor table hides a trick fountain, spurting water from the seats of the benches, the plaything of a prelate who sought amusement as counterpoint to the pomp and circumstance of his stately life (fig. 221). Bomarzo in Italy (fig. 222) is memorable for its Park of the Monsters, which features large grotesque carvings and a topsy-turvy *tempietto*. Even the grand alléed complexity of the Boboli

Billiards-playing monkey carved from a dead beech tree, Bellevue House (see fig. 197)

221

222

223

Fig. 221 Trick fountain at Hellbrunn Palace, Salzburg, Austria

Fig. 222 Grotesques in the Park of the Monsters, Bomarzo, Italy

Fig. 223 Bacchus Fountain in the Boboli Gardens, Florence, Italy

224

225

Gardens (fig. 223), behind the Medici power center of the Pitti Palace in Florence, is refracted through the image of Bacchus, his plump polished-marble belly nestled in a niche near the entry to the garden. These garden images survive, but others may be judged too coy, too opportunistic, or too manipulative. If they are too didactic, the laugh can be forced and awkward; if they are too cute, we blush and turn away. Gnomes need not apply in our garden. But sometimes, the unexpected can be addressed in juxtaposition—by tucking the humor away so that it is not in our face.

Monkey tricks

In our case, we employed monkey sculptures to bring humor and playfulness to the garden. (See chapter 5 for a detailed discussion of the art and craft of these sculptures.) The Chinese saw the monkey as crafty, not a clownish figure aping Western cultural norms.[1] At Bellevue House, monkeys peer across landscapes or peek out from under tables and fountains. Somehow, they add to the amusement of self-discovery; making them less obvious supports the fun. The richness and complexity of presentation dates back to previous centuries. Obviously, the monkeys play both humorous and satirical roles in the Bellevue House garden; some of them don't meet the aforementioned subtlety test, but they do add to the collective drama.

The carving of a monkey in the large stump of a beech tree near the front of the property—one of the huge specimens, now dead or dying, that have added such

Fig. 224 Monkey under the Villa Lante-style table, Bellevue House

Fig. 225 Monkey in swaddling clothes under the children's fountain, Bellevue House

Fig. 226 Drawing of the "Lord" Timothy Dexter Mansion, Newburyport, Massachusetts

226

227

228

229

Fig. 227 *Looking for a New Social Contract* by Justin Gordon, carved out of a beech tree, Bellevue House

Fig. 228 *Monkeys in the Billiard Room* by James Barnes and Deborah Scott-Wilson. This was the inspiration for the *Billiard Buffo* bronze in the grounds of Bellevue House.

Fig. 229 *Billiard Buffo* bronze sculpture with water element by Gregg LeFevre and Jennifer Andrews, Bellevue House

majesty for perhaps 150 years—is meant to bear some weight as political satire. It has an amusing precedent in the wooden statuary that self-styled "Lord" Timothy Dexter erected in front of his Federal-style mansion in Newburyport, Massachusetts (fig. 226).[2] Another monkey looks across the park with a telescope; he has Jean-Jacques Rousseau's social contract under one foot (fig. 227). The sculpture, titled *Looking for a New Social Contract*, was carved by Justin Gordon when Barrack Obama was running for president. He looks across the grounds toward the copper monkey weathervane rotating above the cupola.

The notion of interactive design elements that elicit humor is probably best expressed in the Bellevue House garden in Justin Gordon's carving, in another beech tree trunk, depicting an old monkey, cue in hand, sitting in a billiards chair modeled after one at the Newport Reading Room (a men's club). He is regarding, with a certain melancholy, a bronze bas-relief of four monkeys playing

230

231

Fig. 230 *The Fountain of Insemination* by Mark Mennin. Located next to the lovers' seat, this begins the fountains series at Bellevue House.

Fig. 231 The Lotus fountain by Nicholas Fairplay. The fountain, with its multiplying lotus leaves, symbolizes fertility. It is styled after Edwin Lutyens's design for the Mughal Gardens at the Rashtrapati Bhavan, New Delhi, India.

Fig. 232 This fountain was most likely salvaged from the Arthur Curtiss James estate in Newport by former Bellevue House owner Jane Pickens, who had it assembled. It symbolizes the full force of Western culture with its Renaissance style.

232

billiards that is affixed to the brick wall at the southwestern edge of the property (figs. 228 and 229). *Billiard Buffo* depicts two tough-looking monkeys sporting derby hats and shirt bands with two monkey molls with garters on their legs and décolletage. They are crowded around a billiard table that projects out into space, like the one in Van Gogh's painting *The Night Café*.

The billiard balls are the only fully three-dimensional element in the relief. A foreshortened bourbon bottle, which one of the monkeys has upended while concentrating on making a shot, spills water down the table. The water spurts up as it hits each ball, before trickling into a tank below. An old monkey contemplates this scene from a tree-trunk stump near the house, in another carving titled *Out of the Game.*

A cautionary tale ends a fountain narrative

Billiard Buffo (see fig. 229) can be viewed as part of a sequence of four fountains, a miniature version of the array of splashing jets in Tivoli Gardens at the Villa d'Este south of Rome, though each carries its own message. This sequence is similar to the iconography that might be found in a Renaissance garden, though with a twist. Deepest in the garden is Mark Mennin's huge salvaged classical column that emerges from the topiary just west of the allée. Here, in the shelter of a green-hedged garden room, is the genesis. The lovers come together in the warmth of the sun-soaked stone seat (fig. 230), their act of consummation embracing the classical symbol of the salvaged column.

Next is the lotus fountain set at the edge of a classical pool where the children from the lovers' loins frolic, the lotus symbolizing fertility and beauty rising out of the mud (fig. 231). The main fountain (fig. 232), a shaft of water, represents the zenith of our cultural achievement. It splashes above a carved classical entablature and is on an axis with the architecture of the house. Bellevue House rests on the terrace in view to the north. This principal cultural monument in our cosmos, the plume of water, is lit at night to celebrate our civilizing achievements.

Finally, tucked into the southwest corner of a sheltering wall, is *Billiard Buffo*, where the monkeys who were here first return to take over our cosmos. This fountain iconography can be read as a cautionary tale: if we fail to achieve harmony with our sister nations and with the larger ecology of our natural world, then we are doomed. This appears more probable now than when we inaugurated this sculptural tale some years ago. Perhaps the monkeys are mutants who survived nuclear holocaust or observed our inability to address the global warming crisis that overheated us, and eventually submerged us. This is a grim prospect that perhaps most garden observers won't see. Indeed, many visitors don't spot *Billiard Buffo* and connect the scene with the metaphors of the fountain narrative, and thus don't calculate the international consequences of the game being played.

Conclusion

Thus there is room for humor and satire, for playfulness, in the narrative and the pattern language of a garden. The Bellevue House garden of follies and family experience is not a Disneyland or pastiche, but a deeply considered integration of forms that fit together to shape a larger meaning. Our effort aims to create a townscape of a garden in which the whole becomes more than the sum of its parts. I view this concept of compositional harmony as the ultimate measure of authenticity. Though some preservationists are stuck in the theory of historicism, I believe a mature understanding of the word *contemporary* permits the use of a range of architectural languages; holding to one successive style is a false reading of the notion of "our time."

Preservation bureaucrats often insist on turning historic architecture into a didactic artifact rather than acknowledging its adaptability both as vernacular language and as respected aesthetic composition. Our forced replacement of an existing tapered and belt-corseted Federal-revival-style chimney piece with a modernist stump for the sake of differentiating the different stages of the house's architectural development, as mandated by the Venice Charter (see chapter 1), can now be seen for what it always was, an act of cultural vandalism. We faced similar challenges when attempting to replicate the edging wall along a lazy river at Studley Royal: we failed to understand the essence of the original and instead stupidly substituted clumps of rocks (figs. 233 and 234) for that sublime edge. We only belatedly understood this; our modern landscape designer never did!

Integrity of spirit of place should dictate choice. This is a lesson I've learned as I developed visual literacy, "the conscience of the eye." Architecture is not dead, to paraphrase William Faulkner, it's not even past; it remains a vibrant vocabulary through which our family can reimagine its own sense of place as we extended this vocabulary and celebrate it afresh. Ultimately, I see the expression of harmony as an "authentic" concern that is timeless. It does not depend on a date or a particular architectural period, or the quaint notion that there is a single successor style—Corbu-idiocy.[3] It is all about

233

234

Fig. 233 Rocks clumped at the edge of the pond, Bellevue House

Fig. 234 Pond with corrected edging marked

sustaining a language that is authentic and can be expressed at any time. This sustained authentic pattern language protects our investment in place. This is what we have attempted to do at Bellevue House with our succession of new structures—our cosmos, our little realm.

On the pavement near the Family Chimney Tree and pizza oven, the Fleming family motto is cut into the stone: "Let the Deed Shew," it says (fig. 235). That deed in the ancient and bloody past evidently was the beheading of a clan rival, but we have a new perspective today. Our deed is this creation of a family cosmology in the garden, a narrative facilitated and augmented by the placemaking power of historic associations and the animating strategy of ritual that can similarly engrave memories and associations in the minds of garden users. The integrated art and craft commissions in garden spaces further define them. Finally, the artful crafting of humor and the occasionally satirical garden comment should demonstrate that we don't always take ourselves too seriously, even while orchestrating the harmonics of the garden for what we hope are sublime effects.

The challenge now is for our readers to discover and share their own meanings in their own garden places. I hope we have inspired them with our efforts in building a narrative through memory, placemaking, animation, and deployment of the arts, enhanced by the resonance that comes with recollection and reimagination. I hope our attempts to improve our visual literacy, to stimulate the "conscience of the eye," will further your own endeavors. May your narrative garden strengthen your own family and the families of others who share your adventure.

Notes

1 Alison Waley, *Dear Monkey* (New York: Bobbs-Merrill Company, 1973).

2 The author was executive producer of *A Measure of Change* (1975), a documentary film by Lawrence Rosenblum about the fascinating rebirth of the maritime city of Newburyport, Massachusetts, from urban renewal to civic revitalization.

3 Charles-Édouard Jeanneret (1887–1965), known as Le Corbusier, was a Swiss-French architect and urban planner, and one of the leading proponents of modernism in architecture.

235

Fig. 235 The Fleming Family motto, "Let the deed Shew," in the American Renaissance Water Garden

Artists and artisans

The following artists and artisans were commissioned to provide a level of animation and place meaning to the grounds of Bellevue House. We believe their contributions add to the complexity and richness of the narrative gardens.

Tatyana Murray

Tatyana Murray left the rainy, foggy weather of London and Paris some twenty years ago and was struck by the clear, piercing blue skies of New York City. From that moment, light has been a constant theme throughout her work. Murray fabricated the floating, three-dimensional artwork *Aquatic Dream*, part of her *Light Series*, in the Bellevue House nymphaeum by layering multiple transparent sheets of plexiglass. Through manipulation of the properties of light, temperature, and refraction, the subject matter is elevated to reveal its spiritual essence. Murray creates her unique pieces via a meditative, repetitive pattern-making process in which images evolve and inform themselves. The site chosen for the artwork played a huge factor in the final result. The nymphaeum area at Bellevue House has a discrete opening that leads to a secret back pool/ grotto, and Murray's aquatic-themed light work is the main source of illumination. The glowing images of shoals of fish, seahorses, coral, and seaweed explore an underwater universe in constant flux. At the heart of the disorienting movement lies stillness where the viewer can slow down, reflect, and breathe. Murray's work has been exhibited in New York, London, Paris, Vienna, Barcelona, Cannes, Venice, the Bahamas, and California.

Gregg LeFevre, LeFevre Studios

Since graduating with a degree in philosophy, sculptor and photographer Gregg LeFevre has created over 120 site-specific public artworks in bronze that provide insight about the nature and character of particular places. He often uses cast-metal images and text to illustrate in relief the traits that contribute to the unique personality of a place. His works can be found underfoot in all types of pedestrian spaces, from plazas, parks, bike paths, and trails, to lobbies and arcades. He often works in series, creating a walkway of related pieces. LeFevre's work has been commissioned by governing bodies of New York, Miami, Chicago, Boston, Las Vegas, Seattle, Los Angeles, and many other American cities. He has exhibited widely in the United States. His installations and exhibitions have been reviewed in publications including the *New York Times*, the *Wall Street Journal*, the *Washington Post*, the *Boston Globe*, the *Chicago Tribune*, *American Craft* magazine, and *Landscape Architecture Magazine*. At Bellevue House, LeFevre created the bronze cultural landscape in the Yew Garden, the sculptures in the *Years of Living Dangerously* cascade, the *Billiard Buffo* bronze at the southwest corner of the property, and the Family Chimney Tree in the children's pavilion.

Jennifer Andrews, Masterwork Plaques

Jennifer Andrews received her degree in painting and printmaking from Hartford Art School. She creates work that investigates the structure of memory, myth, and our modern connection to the natural world. Her art has been shown internationally and she has worked as an artist-in-residence in the Netherlands and Vietnam. As part of the public art team Andrews/LeFevre Studios, Jennifer collaborated with natural and cultural historians across the United States to develop illustrated reliefs that are long-lasting, beautiful, and engage the public with a sense of place. In 2014 she co-founded Masterwork Plaques, an artisan plaque and signage company that works with clients such as the American Museum of Natural History, the Smithsonian Institution, the Central Park Conservancy, and the country of Azerbaijan. As Gregg LeFevre's business partner, she was responsible for the outdoor bronze work in the Bellevue House gardens.

Mark Mennin

Mark Mennin graduated with a degree in history from Princeton University, where he also taught ceramics under Toshiko Takaezu. He began to carve stone in Italy in 1984 and worked there for three years before moving to Paris. In recent years, the scale of his sculpture has increased, and he now creates giant landscape and architectural works, often involving hundreds of tons of granite. He has produced a large body of work for both private and public collections, including the fountain and installations at Chelsea Market in New York City; Stanford University; Pennsylvania State University; Bruce Park in Greenwich, Connecticut; Laumeier Sculpture Park; the DeCordova Sculpture Park and Museum; the town of New Harmony, Indiana; Delbarton School in Morristown, New Jersey; SUNY Empire at Staten Island; and the Hotchkiss School, Lakeville, Connecticut. Mennin has taught sculpture and art history for more than twenty years, serving as an adjunct professor at Parsons School of Design and the New York Academy of Art in Manhattan. At Bellevue House, Mennin carved the Villa Lante-style water table, terrace ornamentation, rill, children's fountain, and statue of the goddess Pomona in the American Renaissance Water Garden, as well as the lovers' seat adjacent to the Rosemary Verey pergola and the compass in the Hornbeam allée.

William P. Reimann

William P. Reimann received his MFA from Yale University in 1961, at the height of one of the most fertile and exciting eras at that institution. Working with Josef Albers, Rico Lebrun, and other important artists, Reimann gained early recognition for his extraordinary artistic versatility and technical genius. On the forefront of design techniques in plexiglass and steel, his works were exhibited at—and entered in the permanent collections of—the Whitney Museum of American Art and the Museum of Modern Art in New York before he had completed his graduate work. His art is permanently lodged in the National Gallery in Washington in the J. D. Hatch Collection of American Drawings. Reimann's artwork can be found in numerous public, corporate, and private settings. Most recently, he has been working on intimate, beautifully envisioned works in wood and metal. He crafted the monkey weathervane found atop the cupola folly at Bellevue House.

Nicholas Fairplay

Born in Wakefield, Yorkshire, England, Nicholas Fairplay began his apprenticeship as a stone carver at age sixteen at Chichester Cathedral and thereafter became gargoyle carver on the restoration of the north front and Henry VII Chapel of Westminster Abbey. After receiving his degree in life drawing and clay modeling from the City and Guilds of London Art School, he continued his studies of Renaissance, Baroque, and Roman architecture and sculpture in Rome. Fairplay was recruited as head carver at the cathedral of St. John the Divine in New York, where he directed the carving studio as well as instructing and training an international group of select students and carvers. Upon returning to England, he worked for the National Trust restoring historical buildings and monuments. As a result of Fairplay's extensive knowledge of stone carving, he was invited to be a consultant on the Giza Plateau Mapping Project for the University of Chicago, and the following year he was an expert advisor and was featured in the filming of a Nova documentary on pharaonic pyramid construction for PBS. He is the subject of the documentary *Nicholas Fairplay: Stone Carver,* a co-production of Oberlin College faculty and students and Allen Memorial Art Museum. In 1995, Fairplay was elected a member of the Master Carvers Association of England; he is also a professional member of the Stone Carvers Guild of the United States. At Bellevue House, Fairplay carved the fountain in the Arts and Crafts Garden, which evokes the lotus design that Edwin Lutyens completed at a larger scale for the gardens of the Rashtrapati Bhavan (formerly the Viceregal Lodge) in New Delhi, India.

Luke Randall

Luke Randall has been an active artist in the Rhode Island community for almost thirty years, working professionally as both a decorative painter and a fine art painter. In 1990 he apprenticed with the well-known English decorative painter James Smart. Luke then started his own decorative-painting business called Pompeii Paints. He has worked in many of Newport's grand homes, as well as in Providence, Boston, and New York. He has been studying early American decorative arts for over five years and is a guild member of the Historical Society of Early American Decoration. He has also completed many gilding projects over the years, most recently in New Harmony, Indiana, where he restored the gold leaf on the Jacques Lipchitz gates at Philip Johnson's roofless church. He is interested in classical methods and materials and focuses on realism. A recent painting has been added to the permanent collection at the Newport Art Museum. At Bellevue House, Randall was responsible for the realist murals, faux painting, and gilding in the nymphaeum.

Jeff Buccacio

Jeff Buccacio has worked as a lead designer and special-effects sculptor for many Hollywood motion pictures and is known for his ability to integrate and organize the many aspects of highly complex projects. He crafted the shell bowl and the dolphin and monkey fountainheads in the *Years of Living Dangerously* cascade.

Jonathan Glatt, O&G Studio

Founded by Jonathan Glatt and Sara Ossana, two friends who met as graduate students at the Rhode Island School of Design, O&G Studio has been designing and producing modern heirloom furniture in Rhode Island since 2009. The studio celebrates the rich history of design in New England and continues this legacy, while creating work that is new, surprising, and uniquely its own. O&G has been featured in publications including the *New York Times*, the *Wall Street Journal*, *World of Interiors*, *Architectural Digest*, *Elle Decor*, the *Architect's Newspaper*, *Interior Design*, *Dwell*, and *Martha Stewart Living*, and at the London Design Museum. O&G was named by *Vogue* as one of "ten contemporary designers to collect today." Glatt created the metal arched trellises with repeating keystones found in Hornbeam allée at Bellevue House.

Christa Wilm, Christa's South Seashells

Christa Wilm has been designing and installing a signature line of seashell fantasies for over twenty years. Her undergraduate degree is in journalism, and she has also worked as a writer and producer in Washington, D.C. Later she became fascinated by archeology in graduate courses at Harvard University. She has had a fine antiques store on Antique Row in West Palm Beach, Florida. It was there that she first started experimenting with seashell art. Her work includes furniture and fitted wall panels, as well as ceilings. Christa created the elaborate shell work found inside Bellevue House's nymphaeum.

Clayton Austin, Boston Ornament Company

Clayton Austin of Boston Ornament Company has been doing decorative plaster work for forty years. His company has been involved with design, sculpture, mold making and installation in public buildings, as well as new and restored homes. At Bellevue House he was responsible for the decorative plaster work in the library and nymphaeum.

Nick Tsigaridas

Nick Tsigaridas takes tremendous pride in his work and revels in complicated tile and stone patterns. His Greek family roots have contributed to his expertise today. He fabricated the tile and marble work found in the library and nymphaeum at Bellevue House.

Derek Kosciuszko

Derek Kosciuszko is a woodworker and artist specializing in architectural features, millwork, and boat building. His woodworking skills were integral in creating the exedras, decorative urns, and millwork found at Bellevue House and gardens, as well as the construction of the little teahouse.

Rob Sousa, Modern Concrete

Rob Sousa has been a decorative concrete artisan for twenty-four years. His ongoing mission is to push the limits of concrete. He is most proud of the monument he created at the naval base in Newport to honor the Rhode Island service members killed in the 1983 Beirut bombings. At Bellevue House, he created the decorative grotto stairs in the nymphaeum.

Ndricim Bejko, Boston Art Studio

Ndricim Bejko, president of Boston Art Studio LLC, is a multidisciplinary artist who specializes in monumental sculptures. He created the granite carvings found around the bridge over the *Years of Living Dangerously* cascade.

Emmett Barnacle

Emmett Barnacle was born in Providence, Rhode Island, in 1988. He began blowing glass in high school, before attending the Rhode Island School of Design. After getting his degree from RISD he began his own studio in Pawtucket, Rhode Island, where he creates his cast-glass sculptures. Aside from working on his own pieces, Emmett has worked as a technical assistant for glass artist and sculptor Daniel Clayman since 2014. Barnacle's artistry is found in the custom-made Stalactite Chandelier in the nymphaeum at Bellevue House.

Evelyn Audet, Evelyn Audet Lighting Design

Evelyn Audet, principal designer at Evelyn Audet Lighting Design, located in Rhode Island, has been lighting properties throughout New England and the Eastern Seaboard for over twenty-five years. Understanding how light behaves and the responses it can elicit are crucial to her success. As a lighting designer, Audet uses light as a tool to reveal form, art, details, shadows, color, and texture. Her lighting in the Bellevue House library and rill garden brought forth details of the architecture, interior design, and landscape, enhancing the work of all the other artists and artisans.

Jay Christman

Metalsmith Jay Christman crafted the specialty hinges found on the gates of the Kitchen Garden at the northern edge of the Bellevue House property in the style of master blacksmith Samuel Yellin.

James Barnes and Deborah Scott-Wilson

James Barnes is a painter who has created plein air works for Bellevue House, and Deborah Scott-Wilson is an accomplished ceramicist. Together they created the backsplash behind the stove in the remodeled kitchen in Bellevue House.

Anonymous Artist

The German-born artist who crafted the volutes in the *Years of Living Dangerously* cascade, which catch the water from Jeff Buccacio's bronze monkey and dolphin fountain mounts with incredible precision, has requested to remain anonymous, in keeping with the tradition of the medieval craftsmen who worked on cathedrals; their names were not recorded, but they would sometimes leave their own faces on gargoyles.

Walter Branchi

Walter Branchi is a gardener and a composer. He taught electronic music composition at the Conservatory of Santa Cecilia in Rome and at the Conservatory G. Rossini in Pesaro, as well as in the United States and in Canada. In 1979 Branchi won a Fulbright fellowship to Princeton

University where he first began his major work, the composition *Intero*. In 1983 he was invited by Stanford University as visiting composer at the Center for Computer Research in Music and Acoustics. He founded and directed one of the state-of-the-art electronic music studios in Italy, LEMS (Electronic Studio for Experimental Music). He is the author of the first textbook in Italy on the technology of electronic music: *Tecnologia della musica elettronica*. He worked with UNESCO in Italy publishing theoretical and technical articles and books on contemporary music theory. In 2013 for Open Space New York he published the book *Cantito Infinito: Thinking Music Environmentally*. Mr. Branchi is also an expert in the world of roses, and is the author of *99 anni di rose Tè, Cinesi e Noisette, 1824–1925* (99 Years of Roses: Tea, Chinese and Noisette, 1824–1925). At Bellevue House, Branchi was commissioned to perform the piece *Mysterium Tremendum et Fascinans* within the immense fernleaf beech tree.

Appendices

Appendix 1: Letter from Ronald Lee Fleming to Senator Lamar Alexander, February 2020

Dear Senator Alexander,

I have been to your offices and have seen the extraordinary display of Tennessee history on your wall. It demonstrates that you have the capacity to understand time and place. In my opinion, the vote this week is an essential demarcation of the integrity of our system of government.

In this critical hour, you can be the gentleman who makes the difference.

A few facts:

- I was President of the Young Republicans at Pomona College
- I was in the Ripon Society while at Harvard
- I had an ROTC commission and served in the Fifth Special Forces Group in Nha Trang in Vietnam, and later with the Joint United States Public Affairs Office in Saigon, where I fought for democratic values in the Embassy (and paid a personal price).

I was ready to die for my country and was almost killed by a bullet that flew three inches from my ear.

It is increasingly apparent to Americans that this president has lied repeatedly and is a menace supported by powerful corporate money. I want you to do the right thing and change your vote. You have the perspective to make the difference here and you will be judged historically by that capacity.

Yours sincerely,

Ronald Lee Fleming, Retired Captain
Chairman Emeritus, Scenic America

*We now have a bilaterally supported scenic caucus in Congress, which broadens our reach beyond billboards to redesignating scenic roadways, undergrounding overhead wires, enhancing city entryways, and using health and safety standards to fight digitalization of billboards.

Appendix 2: Draft of a letter by Ronald Lee Fleming, January 1967, Vietnam (see chapter 4, p. 111)

This Christmas letter (draft) ended my JUSPAO days, and when I fought back, ironically sent me to work for "the company."

January 1967

Quite frankly I don't think there is any way we can extricate ourselves though I am pessimistic about the effect of our continued presence here. We are creating a semi urban society which the communists might find easier to organize, but in the process of creating this new society we are not removing the abuses which can make some alternative to the status quo attractive. At this point, in a very pluralistic society the best organized force remains the NLF [National Liberation Front]. So there you are. I agree with most of the things which Halberstam says in that excellent article in November *Harpers* and I specifically agree that we have not learned how to use leverage effectively in our dealings with RVN [Republic of Vietnam]. Furthermore, I think we must eventually realize that this is primarily a political war which, according to General Tra in a captured document located during Operation Junction City, early this year, has two primary objectives—continued instability in Vietnam and increased unrest and discouragement in the United States. If you accept these as their strategic objectives and ignore the U.S. military reports of kill ratios and body counts, then you see that we are not faring very well. This is not to say that I favor our withdrawal; I don't really see how we can. Certainly the success of the sanctuaries in Laos and Cambodia demonstrates that there is something to the domino theory, though I am inclined to place more emphasis on the psychological importance of our stand and its effect on the millions of overseas Chinese who control the economic life of South-East Asia. I do not think that a coalition would work out either, for if we are to accept the political framework of the war, we might as well accept the fact that a form of coalition acceptable to the communists will be only another step in this ruthless political warfare leading to the destruction of the opposition just as it did in North Vietnam as the communists in the Viet Minh wiped out their socialist allies. If I had a solution to this problem, believe me, I would have fired it off in another glittering little memo. Not since the civil war has America ever faced such a crisis of conscience. I hope that we are able to prevail—to find some basis for building a viable political community before we completely destroy this place—but at this point I am not certain that we will be able to achieve that. Certainly the lessons to be learned in Vietnam have not been learned. What is most discouraging I suppose is that some people in government think they have, and that it is only a matter of time before we phase out. Such thinking at this point seems to be unwarranted leap of faith. Perhaps I am not just now philosophically prepared to have faith in anything or anybody, but there is nothing here which inspires me to leap with the rest. If I extend it will be because I think I can make a contribution and because I think that one year service period for officers is a criminal waste of resources.

Please write me if you can make the time; I do appreciate your letters.

Capt. Ronald Lee Fleming
Special Project Officer
Joint United States Public Affairs Office
APO San Francisco 96243

Historic gardens

Studley Royal Park, England
The pond edge at Bellevue House, with its stone steps and the rotunda with cupola situated behind it, evoke a similar arrangement to that of the Georgian water garden at Studley Royal. John Aislabie, followed by his son William, created the Yorkshire masterpiece over more than half a century (1716–after 1768). Aislabie, as chancellor of the exchequer, was a principal sponsor of what was known as the South Sea Bubble scheme. In 1720 when this vast financial operation collapsed, he and his son retired to their garden, which is now a designated World Heritage Site.

The Chinese Dairy at **Woburn Abbey**, seat of the dukes of Bedford, and the scarlet Chinese bridge created in the mid-nineteenth century by John Bateman at **Biddulph Grange**, both in England, directly influence features of the Oriental Vale and the Chippendale bridge at Bellevue House, as did the black beach stones along the pond in the seventeenth-century gardens of the **Katsura Imperial Villa** in Kyoto, Japan.

Villa Lante, Italy
The continuous descending water feature of sixteenth-century Villa Lante in Bagnaia inspire the rill, the stone water table with a trough for cooling wine, and the fountains of the American Renaissance Water Garden at Bellevue House. A succession of two princely cardinals (Gambara, Montalto) owned Villa Lante, which served as a retreat from the Roman summer heat, while also assisting the cardinals in their competition for papal favor.

Little Sparta, Scotland
The sculptures and sometimes unsettling atmosphere of the twentieth-century Scottish garden of Ian Hamilton Finlay and his wife Sue informed the design of the *Years of Living Dangerously* cascade at Bellevue House. The collision of war and nature at Little Sparta (a figure of an aircraft carrier mounted on a column) finds its echo at Bellevue House in sculptures such as a Katanga flag, a hole in the windshield of a Jaguar XKE, a bullet suspended three inches next to a skull, another skull cracked open symbolizing a hemorrhage, and two VW Bugs representing roadside near-disasters.

La Foce, Italy
Features of the pergola designed by the English architect Cecil Pinsent (1884–1963) at Iris Origo's terraced and compartmented gardens at La Foce in Tuscany are referenced in the pergola arrangement adjacent to the Arts and Crafts pool at Bellevue House.

Mughal Gardens, Rashtrapati Bhavan, India
The lotus-pad fountain designed by Edwin Lutyens (1869–1944) for his Mughal Gardens in New Delhi is minutely recreated in the Arts and Crafts pool at Bellevue House.

Hestercombe House, England
The Arts and Crafts pool and garden at Bellevue House, particularly the shell cove at the end of the pool, originate in Edwin Lutyens's early-twentieth-century designs, with Gertrude Jekyll, for the garden at Hestercombe House in Somerset. At Bellevue House, the shell cove is set into a terrace above, and waters at the base flow over into a sapphire pool.

Barnsley House, England
The brilliant yellow laburnum pergola and walk in the Cotswolds garden of famed author and garden designer Rosemary Verey (1918–2001) was created in 1964. Its splendid springtime show is the basis for a similar planting around the pergola at Bellevue House.

Hever Castle, England
Details of the cascade and volute forms in William Waldorf Astor's terrace garden at Hever Castle in Kent helped to inspire both the American Renaissance Water Garden and the *Years of Living Dangerously* cascade adjacent to the library/nymphaeum and emerald pool at Bellevue House.

Het Loo Palace, Netherlands
A sketch by architect and interior designer Ogden Codman Jr. (1863–1951) of a trellis in the gardens of the royal Dutch palace of Het Loo was the direct inspiration for the trellis behind the Pomona statue in the American Renaissance Water Garden.

Iford Manor, England
Garden designer Harold Peto (1954–1933) sketched a bench in Pompeii that he later used in his own gardens at Iford Manor in Wiltshire; that form is echoed in a bench in the American Renaissance Water Garden at Bellevue House.

Bellevue House, United States
The first garden at Bellevue House (originally Berkeley Villa) was a French-style parterre created in the 1920s by the French garden architect Achille Duchêne to complement a summer house designed for Martha Codman Karolik and designed by her cousin Ogden Codman Jr. in 1910.

Schloss Hellbrunn, Austria
The waterspouts incorporated into the Villa Lante-style table at Bellevue House descend from the playful trick fountains devised in the seventeenth century by the humanist and archbishop of Salzburg Markus Sitticus von Hohenems at Schloss Hellbrunn.

Although no features of the gardens listed below were recreated at Bellevue House, they all contributed to the concepts and atmosphere.

Dunmore Park, Scotland

The admirably self-confident Dunmore Pineapple folly stands in Dunmore Park in Stirlingshire. The structure was built during the time of the 4th Earl of Dunmore, John Murray, in the mid- to late-eighteenth century. The design is often attributed to Sir William Chambers, who was known for his whimsical constructions and precise details.

Stowe House, England

Designed by William Kent and built in 1734–35, the Temple of British Worthies at Stowe House in Buckinghamshire is a curving, roofless exedra with niches on either side of a central pier filled with busts. The choice of who was considered a "worthy" was very much influenced by the Whig politics of the family.

Hellfire Caves, England

Cheerful licentiousness characterized the gatherings held in the Hellfire Caves, the meeting place of Sir Francis Dashwood's Order of the Friars of St. Francis of Wycombe, also known as the Hellfire Club. Located in the village of West Wycombe, the caves were excavated from a hill between 1748 and 1752 for Dashwood, founder of the Society of Dilettanti and co-founder of the Hellfire Club. Dashwood and his neighbor at nearby Stowe House, Viscount Cobham, were both members of the Society of Dilettanti, founded in 1734.

Craftsbury Common, United States

Both the design and spirit of this quintessential American town green in northeastern Vermont have inspired the commons area of the Bellevue House garden.

Park of the Monsters, Italy

The Park of the Monsters in Bomarzo, in Viterbo, is a sixteenth-century Mannerist garden populated by grotesque sculptures located among the natural vegetation.

Boboli Gardens, Italy

The Bacchus Fountain is one of the most singular sculptures in the Boboli Gardens in Florence. Completed in 1560 by Valerio Cioli, the fountain depicts the court jester of Duke Cosimo I as the Roman god of wine.

For more information on Bellevue House, please see the following links:

http://www.bellevuehousegardens.com

https://www.instagram.com/bellevue_gardens_newport/

https://alderman-compton-wallpaper-antiques-anew.myshopify.com/blogs/onlybyholly/welcome-to-bellevue-house-gardens

Selected bibliography

Ronald Lee Fleming

Fleming, Ronald Lee. "Adding a Sixth to the House of Five Chimneys: The Case for Aesthetic Harmony over Differentiation." Unpublished manuscript, March 29, 2013.

Fleming, Ronald Lee. *The Art of Placemaking: Interpreting Community through Public Art and Urban Design.* London: Merrell, 2007.

Fleming, Ronald Lee. *Facade Stories: Changing Faces of Main Street Storefronts and How To Care for Them.* New York: Hastings House, 1982.

Fleming, Ronald Lee. "Lovable Objects Challenge the Modern Movement." *Landscape Architecture Magazine* 71, no. 1 (January 1981): 89–92.

Fleming, Ronald Lee, and Melissa Tapper Goldman. "Public Art for the Public." *The Public Interest*, no. 15 (Spring 2005): 55.

Fleming, Ronald Lee, Rachel Goldsmith, and J. A. Chewning. *Saving Face: How Corporate Franchise Design Can Respect Community Identity.* 2nd ed. Chicago: American Planning Association, 2004. 1st ed. 2002.

Fleming, Ronald Lee, and Lauri A. Halderman. *On Common Ground: Caring for Shared Land from Town Common to Urban Park.* Harvard, MA: Harvard Common Press, 1982.

Fleming, Ronald Lee, and Renata von Tscharner. *New Providence: A Changing Cityscape.* San Diego: Harcourt Brace Jovanovich, 1987.

Fleming, Ronald Lee, and Renata von Tscharner. *Place Makers: Creating Public Art That Tells You Where You Are.* 2nd ed. San Diego: Harcourt Brace Jovanovich, 1987. 1st ed. New York: Hastings House, 1981.

Rosenblum, Lawrence, dir. *A Measure of Change.* Executive producer Ronald Lee Fleming. Boston: Urbanimage, Vision, Inc., 1975. VHS.

Newport

Kathrens, Michael C. *Newport Villas: The Revival Styles, 1885–1935.* New York: W. W. Norton & Company, 2009.

Pardee, Bettie Bearden. *Living Newport: Houses, People, Style.* New York and London: Glitterati Incorporated, 2014.

Pardee, Bettie Bearden. *Private Newport: At Home and In the Garden.* New York: Bulfinch, 2004.

Stensrud, Rockwell. *Newport: A Lively Experiment, 1639–1969.* Newport, Rhode Island: Redwood Library and Athenaeum, 2007.

Van Rensselaer, Mrs. John King. *Newport: Our Social Capital.* Philadelphia and London: J. B. Lippincott, 1905.

Architecture

Appleton, Marc. "Civic Architecture in Southern California: Pasadena City Hall and Santa Barbara County Courthouse." *Classicist*, no. 15 (2018): 50–59.

Baker, Paul R. *Stanny: The Gilded Life of Stanford White.* New York: Free Press, 1989.

Cannadine, David, and Jeremy Musson. *The Country House: Past, Present, Future; Great Houses of the British Isles.* New York: Rizzoli, 2018.

Chatfield-Taylor, Adele, and Mary Miers. *American Houses: The Architecture of Fairfax & Sammons.* New York: Rizzoli, 2006.

Clarke, Ethne. *An Infinity of Graces: Cecil Ross Pinsent, an English Architect in the Italian Landscape.* New York: W. W. Norton & Company, 2013.

Cousins, Frank, and Phil M. Riley. *The Wood-Carver of Salem: Samuel McIntire, His Life and Work.* Boston: Little, Brown & Co., 1969. Reprint, New York: AMS Press, 1970.

Curl, James Stevens. *Making Dystopia: The Strange Rise and Survival of Architectural Barbarism.* New York: Oxford University Press, 2018.

Harris, John, and Michael Snodin. *Sir William Chambers: Architect to George III.* New Haven: Yale University Press, 1996.

Hopkins, Andrew, and Gavin Stamp, eds. *Lutyens Abroad.* London: The British School at Rome, 2002.

Irving, Robert Grant. *Indian Summer: Lutyens, Baker and Imperial Delhi.* London and New Haven: Yale University Press, 1981.

Ishikawa, Tadashi. *Imperial Villas of Kyoto: The Katsura and Shugaku-in.* 2nd ed. Tokyo: Kodansha International Press, 1970.

Kimball, Fiske. *Mr. Samuel McIntire, Carver: The Architect of Salem.* Portland, ME: The Southworth-Anthoensen Press, published for the Essex Institute of Salem, 1940.

Labaree, Mabel M., and Benjamin W. Labaree, eds. *Samuel McIntire: A Bicentennial Symposium.* Salem: The Essex Institute, 1957.

Lahikainen, Dean T. *Samuel McIntire: Carving an American Style.* Salem: Peabody Essex Museum, 2007.

Larson, Erik. *The Devil in the White City: Murder, Magic and Madness that Changed America.* New York: Vintage, 2004.

Lessard, Suzannah. *The Architect of Desire: Beauty and Danger in the Stanford White Family.* New York: Random House, 1997.

Lovie, Jonathan, and Sarah Rutherford. *Georgian Garden Buildings.* Oxford: Shire Publications 2012.

Lutyens, Edwin. *Architectural Monographs* 6. New York: Rizzoli, 1979.

Millais, Malcolm. *Le Corbusier, the Dishonest Architect.* Newcastle upon Tyne: Cambridge Scholars Publishing, 2017.

O'Gorman, James F., ed. *Drawing Toward Home: Designs for Domestic Architecture from Historic New England.* Boston: Historic New England, 2010.

Parker, John. H. *The Concise Dictionary of Architectural Terms.* New York: Dover Publications, 2011.

Steil, Lucien, and Leon Krier. *In the Mood for Architecture: Tradition, Modernism and Serendipity.* Berlin: Wasmuth, 2018.

Urban Design Associates. *The Architectural Pattern Book: A Tool for Building Great Neighborhoods.* New York: W. W. Norton & Company, 2004.

Weber, Susan. *William Kent: Designing Georgian Britain.* New Haven: Yale University Press, 2013. Catalogue published in conjunction with the exhibition held at Bard Graduate Center: Decorative Arts, Design History, Material Culture, New York.

Williams, Mathew. *Cardiff Castle.* London: Scala Arts, 2008.

Wilson Richard G. *The Colonial Revival House.* New York: Harry N. Abrams, 2004.

Art and Design

Baert, Barbara. *Nymph: Motif, Phantom, Affect; A Contribution to the Study of Aby Warburg (1866–1929).* Leuven: Peeters, 2014.

Chambers, William. *Designs of Chinese Buildings, Furniture, Dresses, Machines, and Utensils: To Which Is Annexed a Description of Their Temples, Houses, Gardens, &c.* London, 1757.

Codman, Ogden, Jr., and Edith Wharton. *The Decoration of Houses.* New York: W. W. Norton & Company, 1978.

Dannaud, Sylvie, and Gertrude Dordor. *A la cour des singes.* Saint-Remy-en-l'Eau: Éditions Monelle Hayot, 2009. French.

Dwight, Eleanor. *Edith Wharton: An extraordinary life.* New York: Harry N. Abrams, 1994.

Gombrich, Ernst H., and Leonie Gombrich. *The Sense of Order: A Study in the Psychology of Decorative Art.* Wrightsman Lectures 9. 2nd ed. New York: Phaidon, 1984.

Gough, Kathleen M. "Between the Image and Anthropology: Theatrical Lessons from Aby Warburg's 'Nympha.'" *The Drama Review,* August 16, 2012.

Hennessy, John Pope. *Luca Della Robbia.* London: Phaidon, 1980.

Metcalf, Pauline C. *Ogden Codman and the Decoration of Houses.* Boston: The Boston Athenaeum, 1988.

Saarinen, Eliel. *Search for Form: A Fundamental Approach to Art.* New York: Reinhold, 1948.

Selections from the M. and M. Karolik Collection of American Water Colors and Drawings, 1800–1875. Text by Henry P. Rossiter. Boston: Museum of Fine Arts, 1962.

Sweet, Christopher, and Dore Ashton. *Mark Mennin.* Paris: Galerie Enrico Navarra, 2007. Exhibition catalogue.

Gardens and Gardening

Agnelli, Marella. *Gardens of the Italian Villas.* New York: Rizzoli, 1987.

Attlee, Helena. *The Gardens of Portugal.* London: Frances Lincoln, 2008.

Barlow, Nic, and Sally Sample Aall. *Follies and Fantasies: Germany and Austria.* New York: Abrams, 1994.

Bevington, Michael, George Clarke, Tim Knox, and Jonathan Mardsen. *Stowe: The People and the Place.* London: Pavilion Books, published for The National Trust, 2011.

Blanc, Patrick. *The Vertical Garden.* New York: W. W. Norton & Company, 2011.

Brown, Jane. *The English Garden through the 20th Century.* London: Garden Art Press, 1999.

Brunon, Hervé, and Monique Mosser. *L'imaginaire des grottes dans les jardins européens.* Paris: Editions Hazan, 2014. French.

Burdett, Sara. *Biddulph Grange Garden: National Trust Guidebook.* London: The National Trust, 2010.

Campbell, Katie. *Paradise of Exiles: The Anglo-American Gardens of Florence.* London: Frances Lincoln, 2009.

Chambers, William. *A Dissertation on Oriental Gardening.* London, 1772.

Chatfield, Judith: *A Tour of Italian Gardens.* New York: Rizzoli, 1988.

Compton, Tania, and Andrew Lawson. *Dream Gardens: 100 Inspirational Gardens.* London and New York: Merrell, 2009.

Crouzet, François, Michel Saudan, and Sylvia Saudan-Skira. *De folie en folies: La découverte du monde des jardins.* Paris: La Bibliothèque des Arts, 1987. French.

Davidson, Keir. *Woburn Abbey: The Park and Gardens.* London: Pimpernel Press, 2016.

Deitz, Paula. *Of Gardens: Selected Essays.* Philadelphia: Penn Press, 2011.

Envisioning a Greater Perfection: A Garden Conservancy Tribute to Frank Cabot. New York: New York Botanical Garden, 2012. Published in conjunction with the exhibition co-hosted by the New York Botanical Garden and Wave Hill, Monday, April 30, 2012.

Gervais, Paul. *A Garden in Lucca: Finding Paradise in Tuscany.* New York: Hyperion, 2000.

Griswold, Mac, and Eleanor Weller. *The Golden Age of American Gardens: Proud Owners, Private Estates, 1890–1940.* New York: Abrams, 1991.

Headley, Gywn, and Wim Meulenkamp. *Grottoes and Garden Buildings.* London: Aurum Press, 2003.

Heseltine, Anne, and Michael Heseltine. *Thenford: The Creation of an English Garden.* London: Head of Zeus, 2016.

Hitchmough, Wendy. *Arts and Crafts Gardens.* New York: Rizzoli, 1998.

Holmes, Caroline. *The Follies of Europe: Architectural Extravaganzas.* London: The Garden Art Press, 2009.

Howley, James. *The Follies and Garden Buildings of Ireland.* London and New Haven: Yale University Press, 1993.

Jencks, Charles. *The Garden of Cosmic Speculation.* London: Frances Lincoln, 2005.

Jessie, Sheeler. *Little Sparta: The Garden of Ian Hamilton.* London: Frances Lincoln, 2003.

Johnson, Hugh. *The Principles of Gardening: The Classic Guide to the Gardener's Art.* New York: Simon & Schuster, 1997

Jones, Barbara. *Follies and Grottoes.* 2nd ed. London: Constable, 1974.

Kirk, David. *A Year in the Life of Mount Stewart.* Northern Ireland: Cottage Publications, 2012.

Langdon, Philip. *The Private Oasis: The Landscape Architecture and Gardens of Edmund Hollander Design.* Washington, DC: Grayson Publishing, 2012.

Lloyd, Christopher. *Exotic Planting for Adventurous Gardeners.* Cambridge, MA: Timber Press, 2007.

Marston, Pete. *The Book of the Conservatory.* London: Weidenfeld & Nicholson, 1992.

McLeod, Kirsty. *The Best Gardens in Italy: A Traveller's Guide.* London: Frances Lincoln, 2011.

Meers, Nick, and David Wheeler. *Panoramas of English Gardens.* London: Weidenfeld & Nicolson, 1991.

Newhall, Charles W., III. *Brightside Gardens: A Dialogue between the Head and the Heart; The Gardens of Amy and Charles W. Newhall III.* Owings Mills, MD: Bibliotheca Brightside, 2017.

Nichols, Beverly. *Down the Garden Path.* Portland, OR: Timber Press, 2005. First published 1932.

Nitschke, Gunter. *Japanese Gardens.* Cologne: Taschen, 2003.

Notteghem, Patrice, et al. *Les jardins des Duchêne en Europe.* Preface by Michel Baridon. Paris: Association Duchêne; Creusot-Montceau: Écomusée du Creusot; Neuilly: Éditions Spiralinthe, 2000.

Origo, Benedetta, Laurie Olin, John Dixon Hunt, and Morna Livingston. *La Foce: A Garden and Landscape in Tuscany.* Penn Studies in Landscape Architecture. Philadelphia: University of Pennsylvania Press, 2001.

Page, Russell. *The Education of a Gardener.* New York: Random House, 1983.

Pearson, Graham S. *Hidcote: The Garden and Lawrence Johnston.* London: The National Trust, 2007.

Pizzoni, Filippo. *The Garden: A History in Landscape and Art.* New York: Rizzoli, 1999.

Plumptre, George. *Great Gardens, Great Designers.* London: Ward Lock, 1994.

Plumptre, George. *The Water Garden.* London: Thames & Hudson, 1993.

Quinta da Regaleira. Sintra: Cultursintra Foundation, 2013.

Randall, Colvin. *Longwood Gardens: 100 Years of Garden Splendor, 1906–2006.* Kennett Square, Pennsylvania: Longwood Gardens, 2005.

Richardson, Tim, ed. *The Garden Book.* Texts by Barbara Abbs et al. New York: Phaidon, 2000.

Ruggieri, Gianfranco. *Villa Lante.* English ed. Florence: Bonechi-Edizioni, 1983.

Saudan, Michel, and Sylvia Saudan-Skira. *From Folly to Follies: Discovering the World of Gardening.* Cologne: Evergreen (Benedikt Taschen Verlag), 1997.

Sheeler, Jessie. *Little Sparta: The Garden of Ian Hamilton.* London: Frances Lincoln, 2003.

Spencer-Jones, Rae, ed. *1001 Gardens You Must See Before You Die.* Preface by Dr. Elizabeth Scholtz. London: Barron's Educational Series, 2007.

Strong, Roy. *The Artist and the Garden.* New Haven and London: Yale University Press, 2000.

Taylor, Patrick, editor. *The Oxford Companion to the Garden.* Oxford: Oxford University Press, 2006.

Thacker, Christopher. *The History of Gardens.* Berkeley and Los Angeles: University of California Press, 1979.

Tankard, Judith B. *Gardens of the Arts and Crafts Movement.* Portland, OR: Timber Press, 2018.

Triggs, Inigo H. *The Art of Garden Design in Italy.* Atglen, PA: Schiffer Publishing, 2007.

Tunnard, Christopher. *Gardens in the Modern Landscape: A Facsimile of the Revised 1948 Edition.* Philadelphia: University of Pennsylvania Press, 2014.

Verey, Rosemary. *Classic Garden Design.* New York: Congdon and Weed, 1984.

Verey Rosemary. *English Country Gardens.* New York: Henry Holt, 1996.

Wade, Judith. *Italian Gardens.* New York: Rizzoli, 2002.

Watters, Sam, ed. A*merican Gardens, 1890–1930.* New York: Acanthus Press, 2006.

Watters, Sam, ed. *Gardens for a Beautiful America, 1895–1935.* New York: Acanthus Press, 2012.

Whalley, Robin. *The Great Edwardian Gardens of Harold Peto: From the Archives of Country Life.* London: Aurum Press, 2007.

Wharton, Edith. *Italian Villas and Their Gardens.* New York: Rizzoli, 2008. Facsimile of the 1904 ed.

Whitelaw, Jeffery. *Follies.* London: Shire Publications, 2005.

Willis, Peter. *Charles Bridgeman and the English Landscape Garden.* Newcastle upon Tyne: Elysium Press, 2002.

Young, David, and Michio Young. *The Art of the Japanese Garden.* North Clarendon, VT: Tuttle, 2005.

Placemaking

Anderson, Mark. *The Federal Bulldozer: A Critical Analysis of Urban Renewal, 1942–1962.* Cambridge, MA: MIT Press, 1900.

Chernow, Ron. *The Warburgs: The Twentieth-Century Odyssey of a Remarkable Jewish Family.* New York: Random House, 1993.

Cullen, Gordon. *Townscape.* New York: The Architectural Press, 1961.

Duany, Andres, Elizabeth Plater-Zyberk, and Robert Alminana. *New Civic Art: Elements of Town Planning.* New York: Rizzoli, 2003.

Flynt, Anthony. *Wrestling with Moses: How Jane Jacobs Took On New York's Master Builder and Transformed the American City.* New York: Random House, 2011.

Fullilove, Mindy Thompson, M.D. *Root Shock: How Tearing Up City Neighborhoods Hurts America, and What We Can Do About It.* 2nd ed. New York: Ballantine, 2016.

Green, Martin Burgess. *The Problem of Boston: Some Readings in Cultural History.* New York, W. W. Norton & Company, 1962.

Greeves, Lydia. *History and Landscape: The Guide to National Trust Properties in England, Wales, and Northern Ireland.* London: National Trust Enterprises, 2004.

Hobsbawm, Eric and Terence Ranger. *The Invention of Tradition.* Cambridge, UK: Cambridge University Press, 1983.

Jellicoe, Geoffrey, and Susan Jellicoe. *The Landscape of Man: Shaping the Environment from Prehistory to the Present Day.* New York: Van Nostrand Reinhold Company, 1982.

Kemp, Wolfgang. "Visual Narratives, Memory, and the Medieval Esprit du System." In *Images of Memory: On Remembering and Representation,* edited by Susuanne Kucheler and Walter Melion. Washington, DC: Smithsonian Institution Press, 1991.

LaFarge, Albert, and Paul Golderberger. *The Essential William H. Whyte.* New York: Fordham University Press, 2000.

Lessard, Suzannah. *The Absent Hand: Reimagining Our American Landscape.* Berkeley, CA: Counterpoint, 2019.

Nairn, Ian. *The American Landscape: A Critical View.* New York: Random House, 1965.

McWilliams, Wilson Carey. *The Idea of Fraternity in America.* Berkeley,CA: University of California Press, 1974.

Potteiger, Matthew, and Jamie Purinton. *Landscape Narratives: Design Practices for Telling Stories.* New York: John Wiley & Sons, 1998.

Ray, Deborah Wing, and Gloria P. Stewart. *Loyal to the Land: The History of a Greenwich, Connecticut Family.* Phoenix: Phoenix Publications, 1990.

Semes, Steven. *The Future of the Past: A Conservation Ethic for Architecture, Urbanism and Historic Preservation.* New York: W. W. Norton & Company, 2009.

Semes, Steven. "Moving National Preservation Policy Forward: An Opportunity to revise the Guidelines of the National Park Service for Historic Preservation." *Traditional Building,* August 10, 2016; updated January 26, 2017.

Sennett, Richard. *The Conscience of the Eye.* New York: Knopf, 1991.

Storring, Nathan, and Samuel Zipp, eds. *Vital Little Plans: The Short Works of Jane Jacobs.* New York: Random House, 2016.

Tuan, Yi-Fu. *Topophilia: A Study of Environmental Perception, Attitudes, and Values.* New York: Columbia University Press, 1990.

Weyergraf-Serra, Clara, and Martha Buskirk, eds. *The Destruction of "Tilted Arc": Documents.* Cambridge, MA: MIT Press, 1991.

Wilson, William H. *The City Beautiful Movement: Creating the Northern American Landscape.* Baltimore and London: Johns Hopkins University Press, 1994.

Vietnam

Arnett, Peter. *Live from the Battlefield: From Vietnam to Baghdad; 35 Years in the World's War Zones.* New York: Touchstone/Simon & Schuster, 1995.

Boot, Max. *The Road Not Taken: Edward Lansdale and the American Tragedy in Vietnam.* New York: Liveright Publishing Corporation, 2018.

Burrows, Larry. *Larry Burrows: Vietnam.* New York: Knopf, 2002.

Cole, Tom. *Medal of Honor Rag: A Full Length Play in One Act.* New York: Samuel French, 1975.

Halberstram, David. *The Best and the Brightest.* New York: Ballantine, 1993.

Isaacson, Walter, and Evan Thomas. *The Wise Men: Six Friends and the World They Made.* New York: Simon & Schuster, 1986.

MacIntyre, Ben. *A Spy Among Friends: Kim Philby and the Great Betrayal.* New York: Crown, 2014.

Packer, George. *Our Man: Richard Holbrooke and the End of the American Century.* New York: Knopf, 2019.

Sorley, Lewis. *Westmoreland: The General Who Lost Vietnam.* New York: Houghton Mifflin Harcourt, 2011.

Wolff, Tobias. *In Pharaoh's Army: Memories of the Lost War.* New York: Alfred A. Knopf, 1994.

Index

Page numbers in *italics* indicate illustrations and their captions

Adams, Peter Boylston 105–6
Adams, Thomas Boylston 69, 83n.22, 105, 106, 125
adventure playgrounds 65–6
Aislabie, John 19–20, 154
Alexander, Senator Lamar 151
Allen, Lady Marjory (Baroness Allen of Hurtwood) 65–6, 82n.11
Alliance for a Livable Newport 122
Alnwick, Northumberland *136*
American Renaissance Water Garden 30, *38*, *58*, 60n.15, *76*, *89*
  benches 48, *50*, *77*, 90, *91*, 155
  Children's Fountain *24*, *38*, 48, *50*, *58*, *77*, 90, *91*, *130*
    circles representing lives of the children (Mennin) 90, *91*, 144
    monkeys in swaddling clothes 90, *91*, *140*
  Fleming Family Motto 145, *145*
  mortise-and-tenon lintel joints 15, 129, *131*
  rill *38*, 48, *49*, *51*, 70, 88, *89*, 90, *91*, 154
  symbolism and ritual 86–8, 90, 93–4
  trellis *45*, 155
  volutes 48, *49*, 121, *121*, 155
  water table (after Villa Lante) 15, 36, *38*, 48, *49*, *77*, *86*, 86–8, *88*, *89*, 90, *118*, *120*
    author's head carving *15*, *77*, 88, *89*, *118*, *120*
    monkey head carving *130*, 134, *140*
    volutes 48, *49*
    waterspouts *88*, *89*, 90, *118*, *120*, 155
Andrews, Jennifer (Masterwork Plaques)
  monkey relief on the *Billiard Buffo* fountain (with LeFevre) 15, *16*, *141*, 141–2, 143, 146–7
  sculptures for *The Years of Living Dangerously* cascade (with LeFevre) 72, *73*, 81n.1, 113, 122, 146–7
architecture and architectural styles
  artists and artisans involvement 20, 86, 119, 122–4, *123*, 128, 136n.3
  charters and standards 32–4, 35, 60n.4, 143
  classicism 32, 41, 74, 105
  Federal style 21, 25, 27, 32, 34, 35, 37, 41, 90, 141
  modernism 10, 34, 35, *39*, 60n.7, 74, 105, 137n.12, 145n.3
  pattern language 22, 34, 37, 42, 123
  public spaces 20, 82n.14, 93, 137n.12
  see also preservation
Arnett, Peter 117n.5
  Live from the Battlefields 113, 117n.5
Arthur Curtiss James estate, Newport, Rhode Island *142*, 143
artists and artisans
  buildings and gardens 20, 59, 86, 119, 124, 136n.4
  narrative gardens 22, 59, 85, 86
  placemaking 122–4, 136n.9
  public art projects 119, 122–4, *123*, 126–8, *127*, 136n.3, 146
  see also under Bellevue House garden
Arts and Crafts Garden 9, 54, 60n.15, *84*, 90
  Lotus fountain (Fairplay) 8, 52, 74, *76*, 93, *93*, *142*, 143, 147, 154
  pool *51*, 53, *54*, *58*, 74, 90, *91*, 93, *93*, 154, 155
  Reynolds fountain head 52, *52*, 90
  shower arch 42, *47*, 48, 52, 90, *92*, *134*, *135*
Arts and Crafts style 64, 74, 83n.26, 90, *136*
Attingham Park, Shropshire: study program 9, 103
Audet, Evelyn (Evelyn Audet Lighting Design) 149
Austin, Clayton (Boston Ornament Company) 148

Back to the City movement 34
Barnacle, Emmett 149
Barnes, James 149
  Monkeys in the Billiard Room (with Scott-Wilson) *141*
  splashback in Bellevue House kitchen (with Scott-Wilson) *29*
Barnsley House gardens, Cirencester, Gloucestershire (Verey) 54, 75, *78*, 155
Bateman, John 154
Beierle, Mary: Pomona Gate and benches ceramics 124, *124*
Bejko, Ndricim (Boston Art Studio) 149
Bellevue Avenue, Rhode Island 21, 25, 60n.12
Bellevue House, Newport, Rhode Island 7, *26*
  aerial views *8*, *9*, *40*, *41*
  architectural features
    chimneys 35, *39*, 143
    front door with sidelights *133*, 134
    kitchen sink splashback 28, *29*, 149
  classicism 32, 74, 105
  furniture *30*
  history 6, 21, 25, *26*, 27, 28, 30
  interiors
    breakfast room 30, *30*, *31*, 35
    kitchen (new) 28, *29*, *39*
    kitchen (old) 28, *28*
    rotunda and staircase 27, *27*, 55, *133*, 134
  renovation 14, 25, 27, *27*, 28, 30, 32, 35
Bellevue House garden
  "academical village" 22, 37, 81, 101, 109, 112
  celebration of lost companions 22, 81, 101–9, 112, 116
  Library 8, *9*, 36, 41, 42, 48, 65, *70*, 82n.10, *96*, *100*, *101*, *103*, 129, *136*, 148, 149, 155 *see also* nymphaeum *under* garden features *below*
  stable block (conference room and guest bedrooms) 9, 10, 36, 41, 42, *70*, 81, 95, *96*, *100*, *101*, *103*
  aerial views *8*, *9*, *52*, *96*
  artists and artisans involvement 14, 20, 21, 59, 119, 120–2, 124, 129, *130*, 131, *131–3*, 134, 145, 146–50 *see also* named artists
  design influences
    dance and music 22, 54, 93, 150
    elements of architecture 6, 15, *15*, 16, *16*
    family and personal memories 6–7, 17, 20–1, 22, 41, 64, 66, 69, 72, 80, 81, 85, 87–8
    Federal style architecture 28, 35, 41, 42, 48, 53, 82n.10, 90, 101, 141
    humor and playfulness 14, 22, 32, 98, 140, 143, 145
    landmark structures 21, 63, 64, 65, 66, 67, 82n.9
    other gardens and designers 14–15, 16, 22, 56, 59, 73–5, *75–9*, 109, 154–6
    pattern language 9, 21–2, 32, 36, 37, 42, 48, 52–3, 85, 101, 134, 145
    rhythm and repetition 35, 36, 42, 48, 54, 85
  destinations and vistas 22, 30, 32, 41, 52–5, 67, 70, 81, 90
  details
    drains 17, *17*, *50*, 131
    hinges 17, *17*, 22, 142
    mosaics 17, *17*, 20, 32, *51*, 88, *89*, 90, 129, *133*, 134
    trellises 36, *39*, 54, *55*, 60n.12, 101, 129, *131*
    volutes 9, 48, *51*, 53, 70, *79*, 81n.1, 101, 121
  events and activities 98
    association meetings *97*, 98
    concerts 22, *22*, 93, 94
    Daffodillion events 94, *95*, 99n.7

dance 21, 22, 85–6, *86*, 93–4, *94*, *95*, 98, 99n.7
drama 94, 98
garden areas
children's area *13*, 67, 146
common ground (community green) 35, 36–7, 69, *94*, 95, *96*, 97–8
moss garden *9*, 81n.1
public spaces 35, 37, 41, 69, 86, 93
see also named gardens
garden features
bathhouses (cabanas) *9*, 36, 41, 42, 48, 59, 80, *96*, 125
brick columns 30, *31*
brick walls 30, *38*, 60n.12, 142
bridges 41, 48, 70, 88 *see also* Chinese Chippendale bridge
exedras *9*, 42, *42*, *45*, 48, 53, 54, 149
fences and gates 32, 36, 41, *44*, *45*, 48
gazebo (existing feature) *9*, 21, 30, *44*, 53
gazebo (new building) *9*, 42, 48, 53, 54
greenhouse *8*, 36, 41, *50*, *70*, *96*, 131
greenhouse table 131, *132*, 134
lover's seat and enclosure *43*, *142*, 143, 147
nymphaeum *9*, 10, 36, 41, 42, 48, 70, 82n.10, *100*, *101*, *128*, 148, 155 *see also* library *under* "academical village" *above*
nymphaeum grotto 126, *128*, 129, 149
paths and walkways 53–4, 56, 64, 74, *103*
pergolas *2*, 42, *45*, *47*, 53, 54, 75, *78*, 154, 155
pizza oven with chimney 21, 41, 67, 81, 125, *125*, 126, 145 *see also* Family Chimney Tree
pond edging 143, *144*, 154
rotunda with cupola *2*, *8*, 28, 41, *41*, 42, *45*, *46*, 54–5, 73–4, *74*, 80, *96*, 154
stable block *see under* "academical village" *above*
teahouse (one-storey) *8*, 42, *43*, *46*, *47*, 52, *52*, 53, *55*
teahouse (two-storey McIntire replica) 21, *26*, 36, 41, 42, *44*, *46*, 53, 54, 64
see also named garden features
hedges and allées *2*, *9*, 15, *42*, *43*, 48, 60n.15
central allée *52*, 52–3, *53*
teahouse allée *48*
topiary 15, 48
lawns 30, *37*, *43*, 53, 55, 94, 101
materials
beech 41, *41*, 55, *130*, *138*, 140–1, *141*
Brazilian blue quartz 54
brick 30, *31*, *38*, 60n.12, 142
bronze 20–1, 22, 41, *41*, 54, 64, 67, 69, 70, *73*, 81n.1, 119, 122, 125, 129, *131*, 141, *141*, 146–7, 149
glass and plexiglas 70, 119, 129, 146, 149
granite 48, 51, 55, 56, 70, 74, 88, 90, 149
marble 119, 131, *132*, 148
pebbles 17, *17*, *51*, 57, 74, 88
steel 53, 54
wood 53, 54, 55, 64, 82n.10, 101, 119, 131, 149
see beech *above*
as a narrative garden
activities, importance of 67, 85–6
authenticity, importance of 14, 35, 143, 145
family and personal memories 14, 21, 64, 67, 69–70, 80, 81, 85, 90, 125–6, 145
symbolism and ritual 14, 20, 21, 67, 85, 86–8, 90, 93–4, 120, 145
placemaking 14, 20–1, 65, 145
plan (Couture Associates) *11*
sculptures and figures
monkeys 14, 70, 90, 121, 129, *130*, 134, 140, *140*, 142 *see also under* Andrews, Jennifer; Gordon, Justin; LeFevre, Gregg; Reimann, William P.
Neptune 48
putti 90, 134, *134*
representing personal dangers *see Years of Living Dangerously* cascade
representing the Fleming children 52, *52*, 88, *88*, 90, *91*
see also Pomona (goddess)
trees and shrubs *2*, 52–3, 56–7, *57*, 64, 80, 85, *94*
water features 75, *79*
cascades 48, 56, 59 *see Years of Living Dangerously* cascade
fountains 32, 142–3 *see also under* American Renaissance Water Garden; Arts and Crafts Garden; French Garden
pools *9*, 32, 48, 55–6, 70 *see also under* Arts and Crafts Garden; emerald pool
see also Oriental Vale garden
Berkeley Villa/Martha Codman Karolik House (now Bellevue House) 6, 32, 36, 155
Biddulph Grange, Staffordshire: Chinese bridge 56, 154
Boboli Gardens, Florence 139–40
Bacchus Fountain *139*, 140, 156
Buontalenti Grotto *18*
Bois Doré, Newport *30*
Branchi, Walter 22, 149–50
Brolin, Brent 35
Brown, "Capability" 31n.19
Buccacio, Jeff 148, 149
Buckhurst Park, East Sussex 90
Bulfinch, Charles 32
Bundy family 117n.11
Burges, William 20
Burrows, Larry 83n.20, 113–14, *115*
Bute, 3rd Marquess 20

California Club, Los Angeles 82n.18
Calvino, Italo 23n.19
Cambridge Arts Council 126, 127, 128
Campbell, Georgine: portrait of Martha Codman *26*
Cardiff Castle, Wales 20
Chambers, William 56, 74, 129, 131, 156
Chanticleer Garden, Wayne, Pennsylvania 120, *120*
Chihuly, Dale 119
Chinese Chippendale bridge *46*, 56, *57*, 74, *75*, 129, 131, 154
miniature temple bells 74, 131, *131*
Christman, Jay 149
City Beautiful movement 64, 81n.2
Civic Amenities Act (1967), Great Britain 35
Cleveland, Ohio: lamp post *123*
Cobham, Lord 19
Codman, Ogden, Jr. *26*, 27, 28, 36, 60n.11, 102
The Decoration of Houses (with Wharton) 25
drawing of trellis at Het Loo 36, 155
houses designed
Berkeley Villa (Bellevue House) for Martha Codman 25, 27, 32, 53, 155
front gate 36
Land's End for Edith Wharton 36
cove ceiling 30, *30*
Codman Karolik, Martha *26*, 28, 32, 42, 155
collection of American art and furniture 28
former owner of Bellevue House 6, 27, 28, 32
Coleman, Edwin Pope *103*, 104–5
Constable, John Davidson 8, 101–2, *103*
Copley Square plaza, Boston 137n.12
Coral Gables, Florida: palm tree bench *123*
Couture, J. P. (Couture Design Associates) 7, 28
Chinese Chippendale bridge 56, 74, 131
cupola 55, *56*
work in the Library *136*
Couture Design Associates: garden plan *11*
Craftsbury Common, Vermont 9, 95, 97, *97*, 156
Cullen, Thomas Gordon 23n.18
Cumbernauld, Scotland 124
cupola, Branch (or Howard) Street Church, Salem, Massachusetts 54–5, *56*

Daffodillion, Newport, Rhode Island 81n.2, 94, *95*, 99n.7, 122
Dashwood, Sir Francis 19, 156
De Ménonville, Count Thibaut 80
De Tocqueville, Alexis 13

Decanso, Flintridge, California 119
Della Robbia, Andrea: tondo 90, *91*
Della Robbia, Lucca 99n.5
Dennis, John Davidson 12, *103*
Derby, Elias Hasket 32, 35, 36, 42, 60n.2
teahouse designed by McIntire (Derby Summer House) 36, 42, 54, 60n.14
Diodati Dinner Club 11, *97*
Duchêne, Achille 21, 60n.18, 155
French Garden 15, 52, 53
influence on Bellevue House garden 15, 129
mortise-and-tenon lintel joints 15, 129, *131*
Dunmore Park, Stirlingshire, Scotland: The Dunmore Pineapple 19, 21, 156

emerald pool 36, 41, 48, 54, *58*, 59, 60n.15, 70, *86*, *94*
Endicott family 60n.14
English country houses and gardens 9, 19, 23n.7, 23n.11, 36, 114 *see also* named gardens
Euclid v. Ambler ruling, Cleveland, Ohio (1926) 128
European gardens 20, 23n.11 *see also* French gardens; German gardens; Italian gardens

Fairplay, Nicholas 147–8
Lotus fountain 8, 52, *76*, 93, *93*, *142*, 143, 147–8
family and personal memories
garden design influences 6–7, 17, 18, 20–1, 63, 80–1, 85 *see also under* Bellevue House garden
in narrative gardens 20–1, 67, 72, 125
placemaking relationship 6–7, 17, 18, 19, 20–1, 81n.1, 126
Family Chimney Tree 41, 59, 67, 70, 81, 125–6, 145
vignettes 67, *68*, 69, *125*
Fant, Lester Glenn, III ("Ruff") 8, 12, *103*
Finlay, Ian Hamilton 70, 72, *72*, 154
Fleming, Reynolds (son) *22*, *29*, 52, 80, 90, *125*
Fleming, Ronald Lee *98*, *112*
career
townscape design 17, 25, 34, 97, 98
writing on placemaking 7, 8–9, 20, 65, 81n.1, 99n.1, 137n.9
see also military career *below*
childhood 6, 64, 65, 66, 86
collecting 65, 82n.7, 82n.10
"ghost town" 65, 66, 98, *98*
house and neighborhood 65, 66, 82n.12, 97
depicted in sculptures in Bellevue House garden 88, *89*, 90, *118*
education 65, 95, 97, 99n.3, 107, 109, 110, 114, 117n.5
family
ancestors 20, 67, 69, 80, 82n.4, 83n.21, 83n.27, 104, *125*, 125–6
children 21, *22*, 25, *29*, 69, 80, 81n.1, 82n.18, 88, 90, *125*
grandchildren 74, 90, 93, 126
grandparents and great-grandparents 6, 21, 65, 66, 82n.7, 82n.18, 83n.22, *125*
marriage and divorce 25, 69, 102, 104, 105, 126
parents 21, 65, 66, 67, 82n.4
letters 110, 151–3
memories of lost companions 8, 22, 81, 101–9, 112, 116
military career 10, 97, 107, 109, 110, 116
in Vietnam 10, 81, 83n.20, 97, 103, 107–8, 109–14, *112*, 116, 117n.5, 152–3
traveling 9–10, 20, 25, 97, 103, 105
Fleming, Siena (daughter) *22*, 80, 88, 90, *91*, *125*
The Fountain of Insemination (Mennin) *142*, 143
French Garden *42*
Billiard Buffo fountain (LeFevre and Andrews) 15, *16*, *141*, 141–2, 143, 146–7
Codman exedra *45*, 48
Neptune sculpture 48
pollarded trees 48, 52, 53
French gardens 18, 19
Fresh Pond, Cambridge, Massachusetts: drinking fountain 124, *124*
Fullbright, Sergeant Wilson 106–7

Gambara, Cardinal Gianfrancesco 87, 154
Gamble, Jamie 12
gardens and garden design
continuity, importance of 17–18, 63
design influences
family and personal memories 6–7, 17, 18, 20–1, 63, 80–1, 85
humor and playfulness 18, 139, 143
pattern language 22, 32, 42, 63, 85, 143, 145
rituals 93
design styles
eccentricity 18–20
modernism 14, 16–17
garden features
buildings 18–19
follies 18–19, *21*, 156, *156*
grottoes 18, *18*
as public spaces 20, 85
role of artists and artisans 20, 59, 119, 124, 136n.4
symbolism and ritual 17, 85, 122
see also named gardens; narrative gardens
Garland, Hamlin: Middle Border series 104, 117n.2
Gauguin, Paul 72
Glatt, Jonathan (O&G Studio) 54, *55*, 129, 148
Glazer, Professor Nathan 8, 9
Gordon, Justin
monkey as Jean-Jacques Rousseau 41, *41*, 55, 98, 122, 141, *141*
monkey playing billiards *130*, *138*, 141, 142
Graz Art Museum, Austria *33*
Greenwich Common, Connecticut 80, 83n.27

Hadzi, Dimitri: *Omphalos* 126
Halberstam, David 110–11, 114, 117n.11
Harding, David 127
Harries, Mags: *Glove Cycle* 126
Harvard University 97, 102, 104, 106, 109, 114, 117n.5
Graduate School of Design 65, 110
Hellfire Caves, West Wycombe Park, Buckinghamshire 19, *19*, 156
Heseltine, Michael and Anne 20
Hestercombe House and Gardens, Taunton, Somerset (Lutyens and Jekyll) 74, *76*, 90, 155
Het Loo Palace, Apeldoorn, Netherlands: trellis 36, *36*, 155
Hever Castle, Kent *79*, 155
Hicks, David Nightingale 15
Hirschboeck, John 122
Hoare, Henry, II 31n.19, 53
Hoare family 19, 31n.19
Holling, Holling Clancy: *Seabird* and *Tree in the Trail* 64–5
Huet, Christophe: paintings of monkeys 129, *129*, 137n.16
Hult Center for the Performing Arts, Eugene, Oregon: bathroom tiles *127*
Humphrey, Hubert *112*, 113

Iford Manor, Wiltshire 155
Indian gardens *see under* New Delhi
Italian gardens 18 *see also* named gardens
Italian Renaissance 86

Jacobs, Jane 128, 137n.13
Jefferson, Thomas 37, 41, 101
Jekyll, Gertrude 74, 83n.26, 90, 155
Johnson, Lyndon B. 114
Junta (merchants), Salem, Massachusetts 32

Karolik, Maxim 28
Katsura Imperial Villa, Kyoto 57, *57*, *75*, 109
pebbles 14–15, 74, 154

Kennedy, John F. 114, 117n.11
Kent, William 156
Kerry, John 97
Kildare Street Club, Dublin: monkeys playing billiards 15, *16*
Kimball, Fiske 42
King County, Washington 128
Kitchen Garden
details
drain 17, *17*, *50*, *133*, 134
hinges 17, *17*, 149
pebble mosaic 17, *17*, *51*, *133*, 134
gate *45*
herb garden 17, *39*
Knickerbocker Club 117n.9
Korean War 110, 116
Kosciuszko, Derek 56, 149

La Farge, John 136n.4
La Foce gardens, Chianciano Terme, Italy (Origo and Pinsent) 75, *78*, 154
landmarks in American cities 21, 65, 66
removal in urban renewal projects 66–7, 82n.19
Land's End, Newport (Edith Wharton's house) 30, 36
Le Corbusier (Charles-Édouard Jeanneret) 143, 145n.3
LeFevre, Gregg (LeFevre Studios) 146
monkey relief on the *Billiard Buffo* fountain (with Andrews) 15, *16*, *141*, 141–2, 143, 146–7
Pomona College bronze relief *62*, 63–4, 66, 81n.1
replica in Bellevue House garden 64, *64*, 66, 67, 146
sculptures for *The Years of Living Dangerously* cascade (with Andrews) 72, *73*, 81n.1, 113, 122, 146–7
Little Sparta garden, Scotland (Finlay) 14, 70, 72, 154
Longwood Gardens, Pennsylvania 119
Looking for a New Social Contract (Gordon) 41, *41*, 55, 98, 122, 141, *141*
"Lord" Timothy Dexter Mansion, Newburyport, Massachusetts *140*, 141
Louisville, Kentucky: decorative tree guard on Fifth Avenue
Lutyens, Edwin 16, 36, 75, 83n.26, 90
Hestercombe House and Gardens (with Jekyll) 74, 90, 155
influence on Bellevue House garden 8, 52, 74, *76*, 90, *142*, 147–8
Mughal Gardens, Rashtrapati Bhavan 8, 74, *76*, 93, *142*, 147–8, 154
Rajpath with India Gate 90, *92*
Lyman family 59

Malraux Law (1962), France 35
Manini, Luigi 59
Mannerism and gardens 18, 156
Massachusetts Bay Transportation Authority (MBTA): Arts on the Line program 126
McIntire, Samuel 32, 41
arches in Washington Square, Salem Common 42, *47*, 90, *92*, 134, *135*
bathing pavilions for Lyman family 59
carved sofa 131, *132*
influence on Bellevue House garden *2*, 41, 42, 80, 131, *132*, 134
shower arch 42, *47*, 48, 52, 90, *92*, *134*, *135*
two-storey teahouse 21, *26*, 36, 42, *44*, *46*, 53, 54, 64
rotunda with cupola for Branch (or Howard) Street Church, Salem (drawing) 41, *56*, 80
two-storey teahouse built for Derby 36, 42, 54, 60n.14
Mead, Captain Giles 67, 69, 80
A Measure of Change (dir. Rosenblum, 1975) 145n.2
Mennin, Mark 48, 60n.17, 147
blue quartz buttons 54
circles on the Children's Fountain 90, 99n.5, 147
compass 53, *54*
lover's seat and fountain *142*, 143, 147
monkey head carving *130*, 134
statue of Pomona 147
volutes 121, *121*
Messerschmidt, Franz Xaver 134
Model cities program 114
modernism
in architecture 10, 34, 35, *39*, 60n.7, 74, 105, 137n.12, 145n.3
garden design 14, 16–17
preservation relationship 34, 35, 60n.7, 143
Monteiro, António Augusto Carvalho 59
Murray, Tatyana 146
seascape in the nymphaeum grotto *128*, 129, 146
Museum of Fine Arts, Boston 28, 102
My Brilliant Career (dir. Gill Armstrong, 1979) 117n.4

narratives and narrative gardens
activities, importance of 85
artists and artisans involvement 22, 59, 85, 86
authenticity, importance of 14, 35, 143, 145
continuity and compositional integrity 14, 17–18, 35
defined 14
family and personal memories, importance of 20–1, 67, 72, 125
placemaking 14, 20–1, 64–5, 145
in public spaces 126–8, *127*
symbolism and ritual 59, 69, 85, 93
see also under Bellevue House garden
National Museum of Decorative Arts, Buenos Aires: mortice-and-tenon design (Duchêne) 129
New Delhi: buildings and designs by Lutyens 36
Mughal Gardens, Rashtrapati Bhavan 8, 74, *76*, 93, *142*, 147–8, 154
Rajpath with India Gate 90, *92*
New York Botanical Garden 119
Newburyport, Massachusetts 145n.2
document about rebirth 145n.2
"Lord" Timothy Dexter Mansion 141, *141*
Newport, Rhode Island 60n.12, 80, 96, 99n.7, 109, 122
Daffodillion 81n.2, 94, *95*, 99n.7, 122
mansions 21, 25, 26, *26*, *142*
1929 Pierce-Arrow motor car 95, *97*

Oklahoma City 67, 82n.19
Ontario, California: 4th of July picnics 87, *88*
Oriental Vale garden *8*, 23n.13, 56–7, *57*, 59, 60n.15, *75*, 80
lagoon and pool with carp 55, 56, *57*, 59, 73–4, 83n.25
pebbles 57, 74
waterfall 56
see also Chinese Chippendale bridge
Orientalism 57, 59
Origo, Iris 75, *78*, 154
O'Shea, James and John: monkeys playing billiards (Kildare Street Club, Dublin) *16*
Oxtoby, David 137n.13

Page, Russell 23n.11
Park of the Monsters, Bomarzo, Italy 139, 156
Parker, Glenn 7–8
Peabody Essex Museum, Salem, Massachusetts 54
Peel, Fiona 9, 10
Penn, Captain Ronald William 107–9
personal memories and stories *see* family and personal memories
Peto, Harold 16, 75, *91*, 155
Pickens, Jane (former owner of Bellevue House) *26*, 27, 143
Pinsent, Cecil 16, 75, *78*, 154
placemaking
artists and artisans involvement 122–4, 137n.9

family and personal memories 6–7, 17, 18, 19, 20–1, 81n.1, 126
in narrative gardens 14, 20–1, 64–5, 145
in public spaces 122–4
see also under Bellevue House garden
"plop art" 122–3, 128
Pomona (goddess)
relief on the shower arch 90, *92*, 134, *134*
statue (Mennin) 88, *88*, 147, 155
Pomona College, California 63, 66, 87, 95, 107, 109, 137n.13
bronze relief (LeFevre) *62*, 63–4, 66, 81n.1
replica in Bellevue House garden 64, *64*, 66, 67, 146
Pomona College Farm: Pomona Gate and benches (ceramics by Beierle) 124, *124*
Port Eliot, St. Germans, Cornwall 103–4
Portsmouth, New Hampshire 34, 35, *47*
preservation 32–5, 60n.4, 102, 104, 106
conservation areas 35
context, importance of 33, 35
historicism 33, 34, 35, 143
modernism relationship 34, 35, 60n.7, 143
questions of identity 34
Preservation Society of Newport County (PSNC) 122
Proust,Marcel: *Remembrance of Things Past* 63, 81n.1
public spaces
architecture 20, 82n.14, 93, 137n.12
art projects 81n.1, 94, 122–4, *123*, 126–8, *127*, 136n.3, 146
garden design 20, 85
narratives in 126–8, *127*
placemaking 122–4
zoning laws 128

Quinta da Regaleira, Sintra, Portugal 59

Randall, Luke 148
Reimann, William P. 147
monkey weathervane 41, *41*, 55, 98, 120, *121*, 141
E. W. Reynolds Company 66, 69, 82n.4, *125*
Reynolds family 41, 66, 67, 69, 80, 83n.21, 83n.22, *125*
Rosenblum, Lawrence 145n.2
Rousseau, Jean-Jacques 41, 98, 122, 141

Saarinen, Eero 35
Saint-Gaudens, Augustus 83n.22, 119
Salem, Massachusetts 32, 54 *see also under* McIntire, Samuel
Sarah Orne Jewett House, Portsmouth, New Hampshire: pergola (replica built in Bellevue House garden) *47*
Sartre, Jean-Paul 66, 82n.14
Schloss Hellbrunn, Salzburg, Austria: trick fountain 139, *139*, 155
Scottish gardens *see* named gardens
Scott-Wilson, Deborah 149
Monkeys in the Billiard Room (with Barnes) *141*
splashback in Bellevue House kitchen (with Barnes) *29*, 149
Seeger, Pete 109, 116
Semes, Professor Steven 60n.7
Simons, Albert, III: family memories 126
singerie 129, 137n.16
Smiley, Albert K. ("Bert") 66, 82n.16
Smiley Residence, Canyon Crest Park, Redlands, California 66, *67*, 82n.15
Society of Colonial Wars 67
Sousa, Rob (Modern Concrete) 149
St. Germans, Peregrine 9, *103*, 103–4, *104*, 117n.6
Stern, Robert A. M. 35, 63
Stewart House, Count Down, Northern Ireland: Dodo Terrace 20
Stillman, Katherine 10, 105
Stillman, Rufus Chase 105
Stockholm: Kungsträdgården subway 126, *127*
storytelling *see* narratives and narrative gardens
Stourhead, Wiltshire 19, 31n.19, 53
Stowe House, Buckinghamshire: Temple of British Worthies 19, *19*, 156
Studley Royal Park, Yorkshire 14, 20, 23n.13, 55, 73, *74*, 109, 143, 154

Tavern Club, Boston 83n.22, 102, 103, 106, 117n.11, 125
Tempe, Arizona 128
Thenford Gardens and Arboretum, Northamptonshire: rill and pool 20
Townscape Institute 8, 137n.12
Trevelyan, Sir George and Lady 9, 103
Tsigaridas, Nick 148
Tulsa, Oklahoma 67, 82n.19
Tunnard, Christopher 16–17
Tyler, James: *Ten Figures* 126
Tyson family 9

University of Oregon 128
University of Virginia, Charlottesville 37, 41, 101
urban renewal 20, 66, 145n.2

The Vale, Waltham, Massachusetts 59
Vane-Tempest-Stewart, Edith, Marchioness of Londonderry 20
Venice Charter (1964) 32–4, 35, 60n.4
Verey, Rosemary 54, 75, *78*, 155
Vietnam and Vietnamese people 66, 102, *113*, 114
Vietnam War 9, 67, 70, 81, 83n.20, 97, 101–2, 105, 106–8, 109, 110–13, *115*, 117n.5
Villa d'Este, Tivoli: Diana of Ephesus fountain *18*
Villa Lante, Bagnaia, Italy 87, *87*, 154
influence on Bellevue House garden *see under* American Renaissance Water Garden
water features *87*, 154
water table 15, *15*, *77*, 86–7, *87*, 99n.3, 120, 154
volutes 60n.16, *79*, 90, 155 *see also* Bellevue House garden *under* details; *Years of Living Dangerously* cascade

Warner, Charles G. K. "Shot" 102–3, *103*
Warsaw old town, Poland *33*, 34
West Wycombe Park, Buckinghamshire: Hellfire Caves 19, *19*, 156
Wharton, Edith 87
The Decoration of Houses (with Codman) 25
houses
Land's End, Newport *30*, 36
The Mount, Lenox, Massachusetts 36
Wheeler, Scott 122
White, Stanford 119
Whyte, William H., Jr. ("Holly") 128, 137n.12
Wilm, Christa (Christa's South Seashells) 148
Wirtz, Jacques 15
Wisner, Frank, Jr. 116, 117n.9
Woburn Abbey and Gardens, Bedfordshire: Chinese Dairy 56, *56*, 154
Wolfe, Tom 123

Yale University: colleges 35
The Years of Living Dangerously cascade 41, *51*, 70, *70*, *71*, 72, *79*, 117n.6, 121–2, 129, 149, 154, 155
dolphins and monkeys 70, 121, 148, 149
shells 70, 122, 148
skull sculptures (LeFevre and Andrews) 72, *73*, 113, 122, 146–7
volutes *51*, 81n.1, *121*, 121–2, 149
VW beetle sculptures (LeFevre and Andrews) 72, *73*, 81n.1, 122, 146–7
Yew Garden *38*, 48, *52*, 64
Pomona map (LeFevre) 64, *64*, 66, 67

Zorthian, Barry 110, 111

Photography credits

Stephen Barrie / Alamy Stock Photo: fig. 135

Kim Brandell: fig. 185

Gina Broze: fig. 14

Photo by Larry Burrows / The LIFE Picture Collection via Getty Images: fig. 178

Cambridge Arts: fig. 187

Bethy Cardoso: figs. 115, 172, 182

Chronicle / Alamy Stock Photo: fig. 105

JP Couture: figs. 5, 94, 201

Daderot / Wikipedia (CC0 1.0): fig. 16

Bridget Davey – www.bridgetdavey.com, Courtesy The Bedford Estates: fig. 95

Laura DiBiase / iStock Photo: fig. 222

Dinodia Photos / Alamy Stock Photo: fig. 153

European and Japanese Gardens: Papers Read before the American Institute of Architects (Philadelphia: Henry T. Coates & Co., 1902), https://archive.org/details/europeanjapanese01brow: fig. 139

eZeePics Studio / iStock Photo: fig. 140

Ron Fleming Personal Archive: figs. 180, 181

Michael Foley: fig. 9

Fotografo / iStock Photo: fig. 150

Rosalie Frost / Alamy Stock Photo: fig. 121

GardenPics / Alamy Stock Photo: fig. 112

Gionnixxx / iStock Photo: fig. 223

Mick Hales: figs. 8, 10, 11, 12, 13, 24, 25, 32, 33, 34, 35, 36, 37, 44, 45, 46, 47, 54, 55, 56, 59, 60, 63, 65, 71, 72, 76, 78, 79, 81, 82, 83, 84, 85, 92, 96, 108, 109, 111, 113, 114, 119, 126, 134, 136, 143, 145, 156, 174, 180, 183, 190, 194, 195, 198, 202, 203, 204, 208, 209, 210, 211, 212, 213, 214, 215, 220, 224, 225, 230, 232, 233

Chris Hellier / Alamy Stock Photo: fig. 123

Courtesy of Historic New England: figs. 1, 21

ImageBROKER / Alamy Stock Photo: figs. 125, 221

Island Moving Company: figs. 137, 138, 158, 159

Warren Jagger: page 100 and fig. 173

Ernie Janes / Alamy Stock Photo: fig. 131

Nancy Keyes: fig. 168

David Kleyn / Alamy Stock Photo: fig. 192

Robert Kranz: figs. 2, 3, 4, 6, 20, 41, 43, 48, 49, 50, 51, 52, 53, 58, 59, 61, 62, 66, 68, 69, 70, 73, 74, 75, 77, 80, 86, 87, 88, 89, 91, 99, 100, 101, 102, 104, 107, 117, 120, 122, 124, 129, 144, 146, 147, 148, 149, 151, 152, 155, 157, 162, 163, 164, 165, 166, 181, 197, 199, 200, 205, 206, 219, 227, 228, 229, 231, 235

Gregg LeFevre & Jennifer Andrews: figs. 106, 191

Los Angeles Examiner / USC Libraries: fig. 142

MExrix / Alamy Stock Photo: fig. 38

© Max Morriconi, courtesy of La Foce: fig. 129

Courtesy of Museum of Fine Arts, Boston: figs. 22, 207

nobleIMAGES / Alamy Stock Photo: fig. 167

Parker Construction: fig. 110

PCJones / Alamy Stock Photo: fig. 17

Courtesy of Peabody Essex Museum: figs. 93, 154, 218

David Pearson / Alamy Stock Photo: fig. 18

Pomona College: fig. 103

QEDimages / Alamy Stock Photo: fig. 133

Alex Ramsay / Alamy Stock Photo: fig. 98

Kay Ringwood / Alamy Stock Photo: fig. 118

Lisa Roper, Courtesy of Chanticleer Garden: fig. 179

Tomas Sereda / iStock Photo: fig. 39

Geoffrey Smith / Alamy Stock Photo: fig. 116

Francis Specker / Alamy Stock Photo: fig. 127

stevedangers / iStock Photo: fig. 15

ValerioMei / iStock Photo: figs. 7, 141

Waeske / iStock Photo: fig. 40

© Western Morning News / SWNS: fig. 175

All other images are believed to have been provided courtesy of Ronald Lee Fleming and the Townscape Institute.